The Double-Crested Cormorant

The Double-Crested Cormorant

Plight of a Feathered Pariah

Linda R. Wires

With original illustrations by Barry Kent MacKay

Yale

UNIVERSITY PRESS

New Haven and London

Published with assistance from the foundation
established in memory of William McKean Brown.

Yale University Press books may be purchased in
quantity for educational, business, or promotional
use. For information, please e-mail sales.press@
yale.edu (US office) or sales@yaleup.co.uk
(UK office).

Designed by Sonia Shannon.
Set in Fournier type by Newgen North America.
Printed in the United States of America.

Library of Congress Control Number 2013957500

A catalogue record for this book is available from
the British Library.

This paper meets the requirements of ANSI/
NISO Z39.48–1992 (Permanence of Paper).

10 9 8 7 6 5 4 3 2 1

This book is for all the animals
that have been misunderstood

Contents

Preface

THE OJIBWE WORD *NAUBINWAY* (Place of Echoes) is the name of the northernmost and last remaining commercial fishing village on Lake Michigan. Naubinway is also the largest commercial fishing port on the Great Lakes. Fishing is the mainstay of the economy, and freshly caught fish can be bought directly from the docks. Half a mile offshore is a small, two-acre island that bears the same name. Unvegetated and uninhabited by humans, the island is surrounded by rocks and shoals. In 1931, Naubinway Island Light, an unmanned tower aid, was built to guide local shipping traffic. It is the only structure on the island, and though it has been modified, the light retains its original purpose. In the summertime, the tiny island teems with life. Herring gulls and double-crested cormorants begin nesting in May, and by mid-June thousands of chicks are present.

I made my first visit to Naubinway Island in July 1998, unknowingly embarking on a journey that would shape much of my life for years to come. Our crew included Francie Cuthbert, an ornithologist and expert on colonial waterbirds; her husband, Dave Smith, a wildlife biologist; and their ten-year-old son, Charlie. The mild summer day marked my first visit to both a waterbird colony and a Great Lakes island. The purpose of the trip was to obtain blood samples from cormorant chicks hatched at a northern Lake Michigan location as part of a large study by the US Department of Agriculture to determine whether the places of origin of migratory cormorants could be identified. Naubinway Island was chosen for the field visit because it was a northern location and close to a boat launch, always an important consideration when sailing on the lake.

That year, low water levels had made the public boat launch at Naubinway Marina inaccessible, so we drove to the private one used by commercial fishing boats. When we arrived at the dock, two men in a fishing boat watched as we drove up, our truck identifying us in neat letters as affiliated with the University of Minnesota. One man looked to be in his fifties, the other in his thirties; possibly they were father and son. Dave rolled down his window and smiled and asked how they were doing.

"Fine," the older man answered. Neither man smiled back.

"We want to go out to the island there and were wondering if we could use your launch to put in. The water is too shallow at the public boat launch," Dave told them.

"What do you want to do out there?" the man asked.

Francie leaned across toward Dave's window and explained that we were from the university and were conducting research on cormorants. "We want to take a few blood samples from the birds at Naubinway," she told them.

The older man looked from one to the other of us, and then nodded at the island and faintly smiled. "Yeah, those birds are causing lots of trouble. Something needs to be done if we're gonna fish here."

"Fishing bad?" Dave asked.

The younger man snorted while the other sat silent. "Real bad. Won't be no fish left soon. Gotta do something about those birds."

Dave nodded sympathetically.

"Is your research gonna help with that?" the older man asked.

Francie answered that it might help us understand the cormorant population better and this would help understand dynamics in the lake. The man looked skeptical.

"Lake's messed up," the man said, "water's real clear. You can see straight down to the bottom because of those zebra mussels. Between them and the cormorants, you can't make much of a living around here."

After a few more exchanges, the man gave us permission to launch. As we put the boat in, Francie told them we wouldn't be on the island long, no more than half an hour.

The younger man smiled and answered, "Take your time. Get as many cormorants as you can while you're out there. We won't tell."

The journey to the island took ten minutes at high speed. About fifty yards from shore, Dave cut the motor. To minimize disturbing the birds, which are particularly sensitive to intrusions at this time of year, we slowly paddled our way in. The gray water, relatively calm, caused the Zodiac to sway gently from side to side. A strong, acrid smell—the combined guano of thousands of nesting birds and rotting fish—mingled with the fresh moist air. As we came closer, adult cormorants began rising from their nests. They moved first to a standing position and then one after another began vigorously flapping their wings to take off. In seconds, a large, thunderous rush lifted itself into the air, while on the ground the nearly grown but still flightless cormorant chicks crashed through nests, running with wings outstretched, and stampeded through the colony in multiple directions. Many of the gulls, less wary, remained on the island, but called sharply in protest. As we neared, their shrill calls became louder, and some began wheeling overhead. A few feet from the island, Dave jumped from the boat and pulled us ashore. I stepped into the icy water and scrambled carefully over the slippery, algae-covered rocks.

Once on the island, the calls and frenzied movements of the gulls intensified. They flew above us, screaming and occasionally swooping near our heads. Some hundred yards from the island, many adult cormorants had landed in the water. They bobbed on the waves, watching. I focused my eyes on the ground before me. Empty gull nests, spaced casually but liberally apart, occurred along the island's outer edge, while the tall, well-built cormorant nests were densely packed in pods in the center.

Charlie found a large rock upon which Francie set up the blood sampling kit while Dave and I went about retrieving chicks for samples. At the far end of the island, near the water's edge, big cormorant chicks huddled in a large roving mass. As we got closer to the cormorant nests, a mulish braying reached my ears. Half-grown chicks hissed and struck at us with their bills as we passed. Others begged and whined with open mouths. In some of the nests, featherless gray chicks no more than a few days old lay weakly in the sun, a few looking about feebly. Numerous nests still contained only eggs.

I picked up a small, delicate chick, probably a week old, to bring to Francie. At this stage, the large dinosaurian head is unwieldy for the long, sinewy neck. I held the chick carefully while Francie nicked the inside of its wing with a sharp blade. Blood immediately flowed up the small capillary tube. After she was done, I tucked the chick's wing close to its body and returned it to its nest. As we went about retrieving and returning chicks, a gull landed on a cormorant nest near me, picked up a small, naked body with its bill, and flew off. Carrying chicks to and from the rock, I noticed many cormorant nests with broken eggs, pecked by the gulls in search of a meal. Some held large, developed chicks, and some only yolk. We sampled ten birds and spent about twenty minutes in the colony. As we motored away, adult cormorants continued to bob on the waves while the gulls came back to the island in numbers.

Long after our visit to Naubinway, I thought about the tiny island in the vast expanse of water, bare but for its thousands of nesting birds and its one trace of human presence, Naubinway Light. Images of the watchful cormorants bobbing on the water and of gulls swooping over nests lingered in my mind. The journey to the island, though only minutes from the mainland, was a journey to a place forgotten by time but governed by a precise set of rules. Our few minutes in the colony had slanted the fragile balance between the birds in favor of the gulls that day, and many had seized the opportunity for a meal when the adult cormorants were off their nests.

In the desolation of the place and the wildness of its inhabitants, I sensed something indescribable—some sense of knowledge that I had never known or even conceived of. The small spit of rugged land was a world unto itself. And although I had seen cormorants many times before, usually in strong, purposeful flight or on the water, diving in pursuit of fish, I had never really looked at them. The brief experience at Naubinway had an extraordinary awakening effect on me, as though I were seeing cormorants and their world for the first time. In part, their awkwardness on the ground and their vulnerability to the gulls drew me to them emotionally; and that we had come to the island specifically seeking cormorant blood made them stand out in my mind. But something beyond sympathy and the day's mission led my thoughts to return to them again and again.

Over time, I went on to visit many more islands and to study cormorants and other kinds of colonial waterbirds. It slowly dawned on me that the cormorant was quite possibly the most distinctive creature that I had ever encountered. Set apart by appearance, fishing skill, nesting habits, and many other attributes, the cormorant is alone among birds, outside the range of its avian peers. Nesting and roosting in typically uninhabited places, it is mostly a fringe dweller at the edge of human existence. The exception to that separateness, of course, occurs when it fishes. Then it prefers to search for prey in near-shore waters, exploiting opportunities in natural and artificial environments. These waters are likewise highly prized by humans for their fish resources, which brings birds and people into frequent contact. Different from all others and conspicuously successful, the cormorant cannot escape notice.

The same year I made my first visit to Naubinway, events occurred in two other places that would significantly affect cormorants across eastern North America for years to come. The first involved a publication in the *Federal Register*, the official journal of the US government since 1936. On March 4, 1998, the US Fish and Wildlife Service published the Aquaculture Depredation Order for cormorants, the first of two unprecedented policies that allow the killing of large numbers of cormorants without a permit. This order authorizes fish producers at commercial aquaculture facilities (mainly catfish farms) and at federal- and state-owned fish hatcheries to kill an unlimited number of cormorants in order to protect cultured fish. Under this order, birds can be killed only by shooting, but the use of decoys, calls, and lures to bring them into closer range is permitted. The order is in effect in thirteen states, twelve of them in the Southeast, and has primarily targeted cormorants in Alabama, Arkansas, Louisiana, and Mississippi. Hundreds of thousands of cormorants overwinter in the southeastern United States, and many use the high-quality food source found in the abundant catfish ponds. Fry and fingerling fish are stocked at densities of 100,000–250,000 per acre, while the grow-out ponds that raise catfish from fingerling to final harvest are stocked at densities of 4,000–10,000 fish per acre. In 2003, the order was revised to allow for the killing of birds at night roosts in the vicinity of aquaculture facilities from October through April. Through 2010, an estimated 300,000 cormorants had been killed to prevent losses of pond-raised fish since the original order was established.

The second event involved a vigilante-style slaying of cormorants at Little Galloo Island, a forty-five-acre strip of land in Lake Ontario about fifteen miles off the coast of Sackets Harbor, New York. At the time, Little Galloo supported the largest known colony of double-crested cormorants in the world, some 8,500 pairs, and more than 50,000 pairs of nesting gulls. Though cormorants are relative newcomers to Little Galloo, nesting there only since 1974 (as far as we know), the island has long been used by waterbirds, and virtually covered by nesting gulls since 1967. During the last week of July 1998, biologists made a routine visit to the island and encountered a grisly scene: more than 800 dead and decaying cormorants lay before them. A few days later, on August 1, the *New York Times* reported "heaps of carcasses of fledgling cormorants, piles of shotgun shells and starving chicks squawking weakly amid the carnage."

Months after the crime was committed, nine men, several of whom were fishing guides, pleaded guilty to the killing. Angry over poor catches of smallmouth bass that they attributed to cormorants, which they believed to be eating up not just the fish but an entire industry, the men pulled ashore Little Galloo and destroyed as many cormorants as they could. The men were fined and sentenced to house arrest, but eventually the incident was likened to the Boston Tea Party and the vigilantes gained the status of local heroes. This incident brought the cormorant problem into sharp focus for the US Fish and Wildlife Service. The agency took a series of actions that ultimately led in October 2003 to the Public Resource Depredation Order, a second policy that would allow for the killing of tens of thousands of these otherwise federally protected birds yearly. Through 2011, in addition to the birds killed to protect aquaculture, an estimated 146,000 birds had been killed under this second order, and countless numbers of nests and eggs had also been destroyed. Most of these birds were killed at the breeding colonies in the Great Lakes region to protect public resources, such as sport fish and island vegetation.

With these two events lay the start of the modern American war on cormorants. And like many of the wars waged on American predators, the origins to this one have deep roots going back centuries to the arrival of European settlers in North America. Events from that distant time may now seem irrelevant, but upon closer inspection, they are inextricably tied to our present relation to these birds.

In the late 1970s, populations of double-crested cormorants began increasing dramatically across much of their range, and conflicts with humans consequently developed in many areas. In the late 1990s, a group of us at the University of Minnesota were contracted by the US Fish and Wildlife Service to assess the bird's status as a precursor to developing a management plan. While conducting this work, we found that we needed to return to earlier periods to understand the bird's current distribution and abundance. Historical records showed that the cormorant was the owner of a now unimaginable past, and had once existed in many of the places where today it is regarded as a newcomer or an invader, or its numbers are considered unnatural. The importance of historical perspective in understanding the bird's present status became strikingly clear, and so to illuminate its little known past, I decided that this portion of the cormorant's story had to be told in detail.

The cormorant's history, however, was not the only reason for this undertaking. Another, personally more important one revolved around recent management policies and the actions that ensued from them. Not long out of graduate school, with a master's degree in conservation biology, a multi-disciplinary approach that recognizes the need to address the biological and social components of conservation problems, I began working on a textbook conservation biology case: cormorants in the Great Lakes region, one of the continent's hotspots for human-cormorant conflicts. While monitoring birds for the agencies that manage them, I was astonished at the emphasis given to the social dimensions and politics of the cormorant problem, whereas the relevant biology was largely dismissed. Over little more than a decade, I watched the state, tribal, and federal agencies responsible for protecting and managing North American wildlife allow and participate in the destruction of more than half a million cormorants, despite little biological evidence justifying this as a rational course of action. Again and again, decisions were made that lay far outside the realm of any approach to problems I had learned as a conservation biologist, and the cormorant emerged as the victim of many flawed processes and unfair judgments.

More than just an account of a maligned and persecuted animal, the cormorant's story reflects a culture still deeply prejudiced against creatures that exist outside the boundaries of human understanding and acceptance. This cultural bias is the primary impetus for my telling the bird's tale. It is my hope that by revealing aspects of the double-crested cormorant's biology and history, along with the politics shaping its management, some will rethink its role in the ecosystem and its place in human society. And that in learning the story, perhaps some will be motivated to work for real protection and true acceptance of this most interesting of avian citizens.

Acknowledgments

IN THE MID-1990s, I was fortunate to begin working on colonial waterbirds with Francesca J. Cuthbert at the University of Minnesota. She took me to my first cormorant colony and has been a steadfast friend, teacher, colleague, listener, reviewer, and someone to share an adventure with ever since. In the early 2000s, when Ian Nisbet and I discussed undertaking the cormorant's story, he told me that we make time for the things that are important. He was right. He later provided a detailed review of the manuscript and generously looked at several revised chapters. I am indebted to him for his important direction and guidance. Keith Hobson reviewed the manuscript, suggested improvements, and engaged in many discussions with me about this bird. Chip Weseloh and I discussed and argued about cormorants on many occasions, and I thank him for his insights into colonial waterbirds and his answers to my questions about cormorants in Canada. Jim Ludwig freely shared his knowledge of cormorants on the Great Lakes and was always ready to listen and to discuss cormorants. Similarly, Bill Koonz was a fountain of information on cormorants and other waterbirds on Lake Winnipegosis. Ken Stromborg reviewed several chapters, helped me decipher complicated cormorant studies, spent hours with me discussing cormorant policy, offered insights into several aspects of management, and generally kept me inspired to keep going. Activists opposed to cormorant management in Canada, including Barry Kent MacKay, Anna Maria Valastro, Liz White, and Julie Woodyer, provided enormous help with understanding alternative perspectives on events that transpired around culls in Ontario. Their commitment to defending cormorants remains inspirational. I especially thank Barry Kent MacKay, who created and donated the artwork for this book, a gift that brings the text to life in ways that words cannot.

Early on in this undertaking fisheries scientist Walt Lysack provided me with invaluable help in understanding cormorant-fishery interactions on Lake Winnepegosis and elsewhere. In addition, several fishery scientists read sections of this book and provided detailed information about cormorant-fishery interactions. My thanks especially go to Lars Rudstam, Jim Diana, Dave Fielder, Robin DeBruyne, and Andrea McGregor for their help in this area. I did not always accept their criticisms or agree completely with their interpretations, however, and I am solely responsible for the interpretations provided here. I am also grateful to the wildlife biologists Steve Mortensen and Nancy Seefelt for being candid and providing me with detailed observations about cormorant management and fishery science in Minnesota and Michigan, and for their willingness to review sections of this material. Likewise, the avian ecologist Sumner Matteson provided detailed information on cormorants and their management in Wisconsin. The wildlife biologist Steve J. Lewis, who was embroiled in Great Lakes cormorant conflicts years before I came on the scene, has been instrumental in broadening my understanding and study of developments in US cormorant policy and management. The wildlife biologists Patrick Hubert, Pascale Dombrowski, Charles Maisonneuve, and Jean-François Rail helped enlighten me about policy in Ontario and Quebec.

A great deal of the information necessary for this undertaking relied on the early naturalist-writers who left detailed records of their observations of cormorants. I am indebted to the numerous librarians at the University of Minnesota Natural Resources Library and the many graduate students, particularly Dale Trexel, who helped me obtain older works. The comprehensive accounts by Harrison F. Lewis and Howard L. Mendall were especially important and inspired a major portion of this work. The more recent histories of wildlife by Peter Matthiessen and Farley Mowat were also influential.

At Yale University Press, I thank Jean Thomson Black and Sara Hoover for help in making this book a reality. My friends Jane Vercellotti, Jane Hinton, and Lois Baker were good to me during this time. My mother, Bernice, and my brother, Richard, are no longer here to see the book but were part of its making in their own ways. My sister, Susan, has been a lifelong influence, and she read several chapters and provided support during its writing. My daughter, Maria, an ever-present companion, encouraged me throughout the book's undertaking and frequently reminded me that I was "almost there." Finally, I thank my husband, Greg Brosofske, without whom this book would have been impossible and who helped in every way a person could.

PART I

What Are Cormorants?

Great cormorant

1
Aristotle's Raven

An Introduction to Cormorants

TO UNDERSTAND HOW Americans came to find themselves at war with the double-crested cormorant in the twenty-first century, one must first recognize that cormorants are not ordinary birds. To appreciate just how extraordinary they are, consider their occurrence at Disko Bay on the west coast of Greenland in the High Arctic. The coastline is dotted with rocky islands, and glaciers and icebergs abound. In winter, the air temperature can reach minus 22°F or lower, while that of the water, 30°F. Seals, whales, walruses, polar bears, and arctic foxes, all layered with dense fat or fur, traverse the icy waters and frozen ground. The seabirds of the region—gulls, terns, skuas, kittiwakes, puffins, murres, fulmars, eiders—are cloaked in dense feather coats that fully insulate their bodies and protect them from the extreme cold temperatures. All, that is, but for the sleek great cormorant (*Phalacrocorax carbo*), a curious fleck of black that dwells yearlong in this frozen white landscape. Its presence in this polar world is a seeming paradox, for the cormorant originates in the tropics, lacks a substantial fat layer, and, most remarkably, possesses a wettable plumage. The last feature makes it unique among seabirds and a most unlikely candidate for arctic survival.[1]

But not only does the cormorant survive here, it thrives. Between 1998 and 2002, the marine biologist David Grémillet studied the foraging and breeding behavior of cormorants in Disko Bay. He has studied cormorants in many parts of the world, but his work on great cormorants in Greenland provides some especially interesting insights into the adaptability of these birds. Grémillet documented a fish catch rate of 0.6–1.4 ounces of fish per minute, the highest foraging performance recorded for a marine predator. The cormorant's catch per unit effort, a measure of fishing success relative to the amount of effort expended, reached levels ten to thirty times higher than those recorded for other seabird species. In addition, the Greenland cormorants raised on average more than three chicks each season, a substantial reproductive output for a seabird even in a temperate environment.[2]

"One would expect these predatory and reproductive performances to be based upon vast numbers of highly profitable prey within the vicinity of the breeding site," Grémillet writes. Likewise, he tells us, one would predict that the cormorant, essentially tropical and possessing a wettable feather coat, would require disproportionately large amounts of food to compensate for the extensive cooling effect that occurs when it drenches itself in the icy waters. Contrary to these expectations, however, Grémillet discovered that "the estimated abundance of great cormorant prey within their Greenland foraging area was low," and that great cormorants do not have higher energy requirements than other species of diving birds, even when diving in icy waters. How then do cormorants manage to live so successfully in the High Arctic? The answer lies at least partially in the cormorant's exceptional plumage. Unlike most birds that feed on aquatic prey, the body feathers of cormorants are not homogenous. Instead, they have a dual structure: an outer, wettable section and an inner, extremely waterproof section. This feather structure confers two significant advantages: the outer section prevents substantial amounts of air from being trapped in the feather coat, making the cormorant much less buoyant than other birds, and the inner section retains a thin layer of air around the skin and helps reduce heat loss. The result is a bird that is able to dive and actively pursue fish with relative ease in a range of water depths and temperatures.[3]

This delicate balance between buoyancy and insulation is but one of several unique adaptations that enable cormorants to catch fish so efficiently and to thrive in hostile environments. Another may lie in the cormorant's ability to detect and pursue prey. Cormorants are classified as pursuit diving birds, along with penguins, albatrosses, loons, mergansers, and others that dive and pursue fish underwater. This amphibious lifestyle opens a world of feeding opportunities, but also presents major optical challenges because of the markedly distinct properties of air and water. The underwater environment differs significantly from that of the air due to turbidity and luminance. What is more, once submerged the eyes of most terrestrial animals lose their refractive power—the ability to bend light and focus—resulting in farsightedness. But because of their hunting style and fishing success, cormorants are assumed to have excellent visual ability in the water.

From a purely aesthetic viewpoint, the cormorant's eye is often described as one of the most beautiful in the world of birds. Dramatic emerald greens and deep intense blues characterize the eyes of many species, which are all the more stunning and mysterious set against the bird's black plumage. Functionally, cormorant eyes are even more remarkable. Underwater, their unusually flexible lens rapidly adjusts to compensate for the typical loss of refractive power. Additionally, well-developed muscles that regulate pupil size in vertebrates go a step further in cormorants and operate in a sphincter-like manner to further alter the shape of the lens. Studies of the great cormorant indicate that these rapid changes in the lens enable cormorants to attain underwater resolution comparable to that of fishes, seals, sea lions, whales, and dolphins. In some species, a further advantage may arise from a modification of the cornea. In the double-crested cormorant, the cornea is extremely thick and internally flattened, like a penguin's, a pursuit diver that can descend hundreds of yards to pursue fish. Flat corneas, which are associated with lower values of refractive power, reduce the amount of accommodation necessary to achieve a state of emmetropia, perfect vision, when underwater.[4]

Yet just how much the cormorant relies on superior visual ability to detect prey is not clear. Cormorants have been frequently observed to fish successfully in visually limited environments. They are able to forage effectively in highly turbid waters, and in many areas their diet consists of bottom-dwelling and cryptically colored fish. Perhaps most intriguing are more observations from Greenland, where cormorants were studied not only during the breeding season, but also during the winter, when polar night dominates and light levels are very low. Grémillet and colleagues found that rather than changing their daily feeding patterns during this period, cormorants instead made frequent dives in the dark. Such observations led to experimental studies to test the visual acuity of free-swimming great cormorants under a range of viewing conditions. Surprisingly, the results led researchers to conclude that cormorants in fact have poor ability to resolve visual detail in water and are able to detect individual prey only at close range—in fact, less than one yard away.[5]

More recent fieldwork with the Greenland cormorants adds to rather than resolves the mystery of the cormorant's vision. Researchers demonstrated that cormorants forage at shallower depths when light levels are low, and more deeply when light levels are high, indicating that despite their presumed limited vision, cormorants are indeed visually guided. To explain the paradox of how a visually guided predator with poor visual acuity can attain such exceptional fishing success, researchers hypothesize that cormorants may use specialized foraging techniques such as close-quarter prey detection and capture to overcome their visual limitations.[6] So strategy rather than visual prowess may ultimately underlie the cormorant's success. Or, some unknown property of its vision may yet to be discovered. Whatever the reason, the cormorant's amazing underwater performance points to an exceptional adaptation, and is all the more intriguing for having thus far eluded satisfactory explanation.

The Phalacrocoracidae

Worldwide, some thirty-five to forty cormorant species are recognized, depending on which classification is used, and they are found on every continent. During the nineteenth and twentieth centuries, the family, Phalacrocoracidae, was included in the order Pelecaniformes, a diverse group of fish-eating waterbirds that encompassed frigatebirds, boobies, gannets, darters, anhingas, and pelicans. Uniting all are totipalmate feet, a condition in which all four toes are joined by webs; this enhances the ability to swim quickly and strongly. Additionally, each possesses a patch of featherless skin between the branches of the lower bill; in many species, this skin is prominent and forms a pouch for storing fish and other prey while foraging. In pelicans, this feature is developed to the extreme, with the pouch of the American white pelican able to hold almost three gallons of water. But recent analyses indicate that these morphological features have probably converged in birds of different lineages, and pelicans appear more distantly related to these other groups than originally believed. As a result, the cormorant family, along with the frigatebirds, boobies, gannets, and anhinga, have been dissociated from the pelicans and placed in the new order Suliformes, defined in 2010 on the basis of new genetic data.[7]

Most cormorants are similar to a duck or goose in size, although species range from the relatively tiny pygmy cormorant, weighing a little more than a pound, to the very large flightless cormorant of the Galápagos Islands, weighing a robust eleven pounds. During the breeding season, many species

Totipalmate foot of the double-crested cormorant

develop an obvious crest of head feathers, a feature that originally determined whether a species was called a cormorant or a shag. From the Old English *sceacga* and akin to Old Norse *skegg* ("beard," from *skaga*, "to protrude"), *shag* was first used in Great Britain to distinguish between the two species that occurred there, the European shag and the great cormorant. The British form of the great cormorant lacks an obvious crest, and from this distinction a tradition of calling crested species "shags" developed. But the use of either common name is far from consistent. Many species called cormorants have crests, and depending upon the vernacular in use at specific locations, a species can be called a cormorant in one part of the world and a shag in another.

In addition to this confusing common nomenclature, scientists distinguish between two well-defined groups within the family, using the terms *cormorant* and *shag* according to differences in behavior and morphology. The birds called shags inhabit mostly cold waters and are entirely coastal in distribution. They forage offshore, have compact bodies and fair flight ability, and nest on cliffs or on the ground. Those called cormorants occur in mostly warm waters in both inland and coastal environments. The birds forage near shore, are relatively heavy bodied, have more labored flight, and nest on flat ground and in shrubs and trees. Finally, despite numerous biological and morphological criteria proposed to separate cormorants and shags, the terms continue to be used interchangeably for many species, and *cormorant* remains the convenient vernacular for all phalacrocoracids.[8]

All cormorants are darkly camouflaged. Most are black in color, though some are black or gray with a white front, and the Galápagos cormorant is brown. The faces of many are bare, and the facial skin takes on green, blue, or purple tones. All have relatively long necks and bills, along with a significant degree of neck specialization to facilitate bill thrusting. The long neck appears to merge into the head and bill, a feature that has resulted in their being called, along with the anhinga, "snake birds." The edges of the bill are smooth and the bill tip ends in a hook, enhancing the cormorant's ability to grasp and hold fish. With the exception of the Galápagos cormorant, all species have wings of considerable length and width. Together, the sinuous head and neck, substantial wings, and decidedly reptilian facial features suggest more than a vague resemblance to something ancient and pterodactyl.

The North American Cormorants

Six phalacrocoracids are found in North America, and include the double-crested, red-faced, pelagic, Brandt's, Neotropic and great cormorants. All are quite similar in appearance, with mostly black plumage, and to those unfamiliar with the birds, they may appear indistinguishable. Each is an accomplished fisher, yet in North America only the double-crested cormorant is regularly implicated in significant conflicts with humans. What has prevented the other cormorants from becoming the focus of large-scale management and population-reduction programs? In addition to some important differences in social behavior and nesting requirements, the answer lies largely in how and where each occurs on the continent.[9]

The double-crested cormorant is found only in North America, where its distribution is vast. Breeding populations, the unit typically monitored to track population size and trends, occur along the Pacific, Atlantic, and Gulf Coasts, and broadly across the continental interior. The species occupies a great range of aquatic habitats and nests equally well on the ground or in trees. Safety from ground predators and proximity to feeding areas, usually within six miles, are the most important factors driving colony-site selection. Sites meeting these requirements can be found on ponds, swamps, freshwater and saline lakes, reservoirs, lagoons, artificial impoundments, rivers, estuaries, and open coastlines. Cormorants prefer to locate colonies on islands, anything from small boulders and sandy shoals to larger, densely forested landmasses. Cliffs are also used, as well as artificial structures such as bridges, towers, and navigation aids. Occasionally, colonies occur on portions of the mainland that are free from predators.

In the winter, many birds move south from northern breeding areas to find open water and feeding, loafing, and roosting sites. Birds breeding along the Pacific Coast are mostly resident, and the entire coast from Alaska to Sinaloa, Mexico, remains occupied. In the East, inland and coastal portions of states from Pennsylvania to Florida are occupied, and the mid-South constitutes the major wintering area, where the majority of birds are found along the Gulf and southern Atlantic Coasts. Concentrations gather both inland and offshore on barrier islands. Much of the wintering habitat for interior breeding birds is concentrated near the mouths of major river systems, including the Mississippi, Arkansas, and Red Rivers. Extending linearly with these rivers, wintering habitat radiates toward northern breeding areas and provides well-defined migration pathways. Band returns indicate that birds migrate to the Lower Mississippi River Valley from an area extending from Alberta to coastal New England, and everywhere in between. This wide distribution, combined with the ability to exploit a diverse array of habitats, inevitably leads to frequent encounters with humans, which have gained this noncosmopolitan species a global profile.[10]

By comparison, North America's other cormorant species are far more limited in distribution. Red-faced, pelagic, and Brandt's cormorants are encountered only along the Pacific Coast and have much more restrictive habits and requirements. The Neotropic, the only tropical cormorant to occupy the continent, is quite versatile in its use of habitat, but the bulk of its range is in Central and South America. In North America, it occurs only from southern New Mexico to southwestern Louisiana and down through Mexico. Finally, the great cormorant is found in a diversity of interior and coastal habitats; it is the most widely distributed of all suliform birds. In North America, however, it is restricted to the north Atlantic Coast, where it nests on sea cliffs and rocky islands in small numbers. But in Europe, it has a continental distribution, and conflicts with fisheries are at a pan-European scale, mirroring those of the double-crested cormorant in North America.

The Definitive Fisher

Although cormorants have many fascinating attributes, it is the bird's remarkable skill as a fisher that has captured the human imagination and that defines it for most people familiar with this group of birds. Within the family, there is a great diversity of fishing styles and techniques, and the talent exhibited by a few in particular has made the group of great interest to humans for millennia. Fishermen began exploiting the skill of great cormorants by taming individual birds and teaching them to bring back fish more than two thousand years ago; the earliest known record of fishing with cormorants is from China in 317 BC. This practice involves a great deal of care and continual training of the birds, which are raised from hatching so that the first thing the chick sees is his human partner. The birds are taught to catch fish for their handlers and to return them to the boat, eventually becoming an indispensable aid to the fisherman's trade. Some birds are so highly regarded that when it is time to pair them, birds are allowed to choose their own mates so that the pairing is a "love match" and the birds find happiness. This practice is still used today in continental China by commercial fishing operations. On the Li-Kiang River downstream from Guilin, it is stated that a cormorant fisherman can readily earn more than five times as much as a university professor.[11]

The Japanese have a long, rich tradition of fishing with both great and Japanese cormorants; they reportedly engaged in nighttime fishing with cormorants as early as the eighth century AD. This custom became a highly stylized court event early on, and though it has been replaced as a commercial fishing venture by more modern methods, it is protected under the Imperial Household Agency and still practiced in order to preserve the tradition and attract tourists. The Japanese report that efficient cormorants are able to catch up to 150 fish an hour. In Europe, fishing with great cormorants was practiced in England, France, and Belgium in the sixteenth and seventeenth centuries, though it was done more for sport, like falconry, than for commercial fishing.[12]

Interest in cormorant fishing is perhaps more widespread now than ever before. Over the last century, hundreds of scientific papers have been published describing cormorants' fishing preferences, behaviors, and impacts on fish populations on all continents. This interest arises not so much out of curiosity as out of practicality, and is driven mostly by just a few species. In Europe and North America, a seemingly infinite number of stories related to cormorant fishing have been covered by the media, whose role in the double-crested cormorant's story could easily fill a book. Most recently in the United States, millions of dollars have been spent to reduce the number of fish eaten by cormorants in the southern and northern states (see chapter 10).

Highly efficient in even the harshest environments, the cormorant has been aptly described as "a true athlete" by David Grémillet. But alas, the bird's amazing talent is a double-edged sword, for it is precisely this talent that brings it into direct competition with humans and predisposes it to conflict with our ever-more territorial species. As a result, cormorant species have been persecuted on several continents. And thus the ultimate paradox associated with the cormorant is that its athleticism and effectiveness are the attributes that most threaten its ability to survive today.

The Fossil Record and the Logic of Form

A sense of history should be the most precious gift of science and the arts, but I suspect that the grebe, who has neither, knows more history than we do. His dim primordial brain knows nothing of who won the Battle of Hastings, but it seems to sense who won the battle of time. If the race of men were as old as the race of grebes, we might better grasp the import of his call. Think what traditions, prides, disdains and wisdoms even a few self-conscious generations bring to us! What pride of continuity, then, impels this bird, who was a grebe eons before there was a man.

—Aldo Leopold, *Sand County Almanac* (1949)

Although it is a matter of some debate exactly when the first bird originated, by the time such prehistoric superstars as *Tyrannosaurus rex* came to dominate the large-predator niche in North America, and *Triceratops* was an abundant prey item, orders of birds had already existed for tens of millions of years. Some paleontologists have located the avian starting point with a crow-sized, birdlike creature called *Protoavis*, whose fossil remains date back 225 million years. But the remains are fragmentary and lack direct evidence of the all-important feather. Most avian paleontologists, therefore, identify the beginning of birds with *Archaeopteryx*, for which there are beautifully preserved fossil specimens showing an aerodynamic wing and detailed impressions of feathers. Crow-sized and distinctly birdlike, *Archaeopteryx* hails from

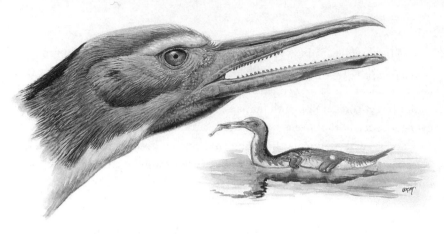

Hesperornis

the Late Jurassic and the age of dinosaurs, indicating that birds are at least some 150 million years old.

Almost as ancient as *Archaeopteryx* but less well known is *Enaliornis*, for which there is a fossil specimen dating back to the Early Cretaceous, the period following the Late Jurassic, 100 million–135 million years ago. *Enaliornis* is the oldest known specimen of the hesperornithiforms, an order of foot-propelled diving seabirds that ate fish and were almost entirely aquatic. Fossil specimens retrieved from the Cretaceous, a period spanning 80 million years, indicate a very successful order with multiple genera and species and a worldwide distribution.

These early divers disappeared some 80 million years ago at the end of the Cretaceous, about the same time as the dinosaurs. But their highly functional form had an obvious logic, and specialized foot-propelled diving birds subsequently arose again several different times from independent lineages. Re-creations of several hesperornithiforms based on fossil remains depict birds that look remarkably similar to their modern but unrelated counterparts, the foot-propelled diving loons, grebes, and cormorants. The ancient birds possessed diminutive wings and were flightless, but had large feet and were well adapted for diving and swimming. They may have nested in colonies on isolated coastlines or islands in the Cretaceous seaways. They would have been clumsy on land, and probably could not walk upright, characteristics that have led some to speculate that they may never have ventured onto dry land at all but instead gave birth to live young. Fossil remains of the order have been collected from several North American locations, including many important fossils from the interior of the continent. Today, the flightless Galápagos cormorant, with its tiny wings and heavy mass, has the highest wing-loading ratio of all the cormorants and has been identified by the biologist Colin Tudge as the nearest thing among modern seabirds to this long-gone order of foot-propelled divers.[13]

The earliest traces of the Suliform birds, based on a frigatebird fossil, are from the lower Eocene, approximately 56 million years ago. Definitive fossils identified as Phalacrocoracidae date to the Eocene-Oligocene boundary, indicating that cormorants have been present for at least 34 million years. Because Australasia has more species and more representatives of various subgenera, its archipelagoes and the former inland seas of Australia have been proposed as the potential points of cormorant origin, divergence, and radiation. But more recent work has indicated that additional data are needed in order to evaluate global hypotheses about the bird's origins.[14]

By the time true grebes appear in the fossil record, roughly 25 million years ago, in the late Oligocene or early Miocene, much of Eurasia was covered by shallow seas, and the first elephants with trunks, early horses, and many grasses had appeared. Additionally, cormorants had diverged from their closest ancestors, the anhingas, and several representatives from the modern genus *Phalacrocorax* are identifiable. Fossil bones of one of the earliest known species, *mediterraneus*, were retrieved from the White River formation in Weld County, Colorado, and date back to the lower or middle Oligocene. Remains of numerous other species dating to the Miocene were found in California, Oregon, and Florida, as well as in France and Germany.[15]

Several species resembling or overlapping in measurements with the double-crested cormorant are known from the Pliocene, an epoch extending from 5.3 million to 2.6 million years ago. Fossils of these birds have been retrieved from California, Idaho, and Florida, locations where double-crested cormorants occur today. Bones definitively belonging to the double-crested cormorant itself date back to the most recent ice age in the Pleistocene, a period spanning 2.6 million to 12,000 years ago. During this time, the double-crested cormorant was widespread, just as it is now; fossils have been retrieved from the East and West Coasts of North America and from interior locations. Additionally, archeological evidence from middens 500–5,000 years old indicate that double-crested cormorants were breeding on both the West and East Coasts in approximately the same areas where they still breed today.[16]

By comparison, while the first primate-like creatures began evolving approximately 60 million years ago, nothing remotely resembling humans appears in the fossil record until relatively recently. The earliest known specimen of the family Hominidae, to which humans belong, is from a skull discovered in 2001 by the French paleontologist Michel Brunet and colleagues; it dates back some six million to seven million years ago. Research and virtual reconstruction of the cranium confirmed that this creature was a hominid, but neither ape nor human. Arguably, the oldest known representatives of the genus *Homo*, in which humans are classified, are the species *habilis* and *rudolfensis*, creatures that evolved an estimated 1.9 million to 2.5 million years ago and still retained some apelike features. Modern humans, *Homo sapiens*, emerged roughly 200,000 years ago, and today are the sole representatives of the genus *Homo*.[17]

While the evolution of cormorants and humans both occurred in just the last few minutes that have ticked by on the geologic clock, the species are vastly separated by time. The cormorant is a comparatively ancient creature, distinct some 30 million years before humans appeared. Its origins call forth a remote primordial universe that has seen a multitude of evolutionary forms come and go. Nevertheless, from *Enaliornis* to *Phalacrocorax*, foot-propelled diving birds have remained. Repeating and refining adaptations that were highly adaptive tens of millions of years ago, theirs is an essentially relic form perfected by the ages. During its millions of years of existence, the cormorant has not changed greatly in size or appearance. And for most of its time on earth, it has been undisturbed by humans, for, as Leopold observed of the grebe, the cormorant was a cormorant eons before there was a man. In this context, then, it is a truly ancient continuity that impels the cormorant ever onward in its customs and traditions.

Aristotle's Raven: What's in a Name?

The most obvious way of tracing an idea is to go directly to the oldest source and see whether it occurs there. If it doesn't, the alternative is to work back through the literature. This is not as easy as it sounds, and is rather like trying to trace the source of a river: as the stream becomes smaller the terrain becomes more rugged,

and the stream often divides into smaller and smaller rivulets,
making identification of the true source a challenge.

—Tim Birkhead, *The Wisdom of Birds* (2008)

Early efforts to understand, classify, and distinguish cormorants relative
to other birds date back to the writings of Aristotle, who around 350 BC composed *History of Animals*, one of the most influential works in zoology and
natural history ever completed. This ten-volume opus—the first real effort in
Western culture to classify living things—describes more than five hundred
animals and identifies important relationships and differences across the animal
kingdom. For the many birds described in book 8, Aristotle developed broad
categories based on their occurrence on land or on water, and subcategories
based on differences in diet and foot anatomy. For waterbirds, two groups were
distinguished: those that were web-footed and lived on the water, and those
that were lobe-footed (toes separate but each surrounded by a fleshy lobe) and
lived by the shore. The web-footed group comprised several large kinds of
waterbirds that live on the banks of rivers and lakes, including "the swan, the
duck, the coot, the grebe, the teal . . . and the water-raven [*hydrokorax*]." The
last is described in detail: "This bird is the size of a stork, only that its legs are
shorter; it is web-footed and is a good swimmer; its plumage is black. It roosts
on trees, and is the only one of all such birds as these that is found to build its
nest in a tree." The "water-raven" is of course the cormorant, the only large
black waterbird with a webbed foot that builds its nests in trees.

To the modern reader, Aristotle's name for the bird may seem a curious
choice, given that the cormorant is not a raven at all. And certainly Aristotle
was aware of this, given his keen observational skills and detailed treatment of
ravens elsewhere. But in ancient Greece, precise nomenclatures for birds did
not exist. Rather, writers in antiquity often grouped many different species
under one name, and writers from different periods or birthplaces often called
the same birds by different names. In *Birds in the Ancient World from A to Z*
(2007), the classical scholar W. G. Arnott identified two birds that shared the
name *korax* in ancient Greece: most frequently the raven, named for the sound
of its coarse, unmusical call, and second, the cormorant. In addition to these,
other birds were identified as types of *korax*, such as the nocturnal *nyctikorax*,
"night-raven," which included owls and night-herons, and the *rhinokorax*, the

"raven with a prominent nose," which probably identified the western reef egret, a crepuscular bird loosely resembling a vulture.

Common features linking these diverse birds were probably obvious characteristics related to their appearance, behavior, and habits. In ancient Greece, these external features expressed essential qualities inherent to particular kinds of birds, and were significant because they conveyed meaningful information to humans. In his comprehensive work *Birds in Greek Life and Myth* (1977), the classical scholar John Pollard describes the importance of birds in the nearly universal Greek belief in omens. Special meanings were inferred, and messages communicated, by almost any action natural to a bird, including vocalizations, wing-flaps, and night flights. Birds were able to divine signs given by the gods, serving as their agents and heralding things to come. This belief, coupled with practical observations on bird behavior, presumably led to the development of augury, "the reading of signs," whose practitioners in ancient Greece were the *oionistai, oionopoloi,* or *oionomanteis,* "experts in bird omens." Certain birds were considered more ominous than others, and the raven, according to Pollard, with its "somber magnificence, croaking call and great intelligence was not surprisingly regarded as highly ominous."[18] Familiar birds in the towns and cities of ancient Greece, ravens were admired for their ability to mimic and entertain. But they were also of evil repute, considered "guilty of sacrilege" for scavenging the dead and robbing sacrificial altars. The similarly sized cormorant, with its all-black plumage, croaking call, and decidedly somber presence, was an obvious candidate to be the raven's aquatic counterpart. But the cormorant was a bird of seacoasts, lakes, and rivers, and although frequently encountered by fishermen and sailors, it did not inhabit the landscape of human ritual and custom to the same extent as the raven. Nevertheless, its physical similarity and its identification as a type of raven suggest that the cormorant probably was perceived to possess some of the same potent qualities that gave the raven its iconic status.

For the next two millennia, Aristotle's "water raven" shaped the cormorant's identity. The unique name appears approximately four hundred years later in Pliny the Elder's *Naturalis Historia*, a landmark encyclopedia in Roman culture consisting of thirty-seven books. One of the most influential thinkers and writers known from ancient Rome, Pliny treated a vast expanse of topics and frequently drew upon Aristotle. In book 11, chapter 47, Pliny provides an anatomical discourse on baldness. Writing of lands and soils where people have no hair at all, Pliny states, "Some kinds of animals are bald by nature, for

instance ostriches and cormorants (*corvi aquatici*)." Pliny adds that it was the
water raven's baldness that led to its Greek name, *phalacrocoraces* (Latinized
Greek from *phalakros*, "bald," and *korax*, "raven"), which is given in book 10,
chapter 68. Pliny notes that phalacrocoraces are found in the Alps and are
"specially belonging to the Balearic Islands." Not surprisingly, later naturalists concluded that the bald *corvi aquatici* of book 11 and the *phalacrocoraces* of
book 10 were one and the same, and that Pliny was referring to the cormorant
when he used Aristotle's "water raven" to exemplify creatures that are bald.

Pliny's comments sealed the cormorant's identity but provide little description of the *corvi aquatici*, other than that some are bald. Likewise, the only
distinguishing feature given for the *phalacrocoraces* is their apparent baldness.
But cormorants, in fact, are not bald. Some species develop fine white feathers on their heads and necks during the breeding season, a feature that can
give the appearance of baldness; but Pliny describes a baldness that involves a
lack of feathers. Therefore, some scholars believe that Pliny's *corvus aquaticus*
and *phalacrocoraces* refer to an altogether different bird; Arnott suggests it was
probably the bald ibis, a superficially similar black waterbird that, like the vulture, lacks head feathers entirely.

The imprecise nomenclature that ancient authors used may forever preclude our ability to identify definitively many of the birds they described. But
in tracing the cormorant's identity, later translations appear equally important, regardless of their accuracy. Some four hundred to five hundred years
after Pliny's treatment, Hesychius of Alexandria compiled a large dictionary
of classical Greek words, the most important one to survive the Middle Ages.
Hesychius translated *aithyia*, another seabird described by Aristotle and translated by Pliny as *mergus*, which could refer to a number of marine diving birds,
as the marine *korone*, or "sea crow." In ancient Greece, *korone* commonly identified the hooded crow, but Arnott pointed out that *korone* could also refer to
other birds, including cormorants and shags, "crowlike birds more attached to
the water than the Hooded Crow."[19]

Whatever the original identity of Aristotle's *aithyia*, the notion of a "sea
crow" traces the cormorant's identity from the classical world through the
Middle Ages. In the eighth century, the essentially analogous Latin term *corvus
marinus*, "sea raven," was documented in the Latin-Reichenau glosses to the
Vulgate Bible and translated as *mergulus*, a version of *mergus*, directly linking
the classical and medieval names for this bird.[20] *Corvus marinus* was eventually
contracted to *cormoran* in Old French and to *cormeraunt* in Middle English.

The *Oxford English Dictionary* traces the first use of *cormorant* to the end of the Middle Ages, circa 1320.

Sometime after the bird became known as the cormorant, its association with the raven family began eroding in popular culture, but persisted among those interested in formal names for birds. In 1544, the English natural historian William Turner published *Turner on Birds: A Short and Succinct History of the Principal Birds Noticed by Pliny and Aristotle*, in which he translated roughly 110 bird descriptions made by his predecessors. For the cormorant, he attributed the more correct use of *corvus aquaticus* and *mergus* to the Swiss naturalist Conrad Gessner, noting, "The Corvorant is a voracious bird. . . . Our people corruptly say Cormorant, not knowing from the derivation of the word that it ought to be called the Crow that devours." About a decade later, the French naturalist Pierre Belon du Mans published *Portraits d'oyseaux*, which included the names by which the cormorant was known in Latin, Greek, French, and Italian. All but one was a version of "water raven." In 1570, the British scholar and physician John Kay Caius published *Of Some Rare Plants and Animals*, in which he chided people for using the name "cormorant," stating that "corvorant," deriving from *corvus marinus*, or water crow, was correct.

About one hundred years later, the English naturalist John Ray published *Willughby's Ornithology*, the product of a collaboration with his friend and student Francis Willughby, who died before the work could be published. Like Aristotle, Ray and Willughby divided birds into land and water species, but further subdivided them according to the shape of their beaks and feet. They classified ravens as land birds with straight and large beaks, and cormorants as waterbirds that swim and have four webbed toes. Nevertheless, they perpetuated the cormorant's raven association, maintaining the name *Corvus aquaticus* for the great cormorant and introducing *Graculus palmipes* for the European shag, the latter translating as "web-footed" (*palmipes*) "jackdaw" (*graculus*), the smallest member of the raven family.

In the eighteenth century, the Swedish physician and naturalist Carolus Linnaeus published *Systema Naturae*, which became the universally accepted system of scientific classification and naming. This system identified species by utilizing a unique two-part name chosen to be descriptive and to provide information about an organism's relationships to other organisms. In the tenth edition (1758), Linnaeus classified cormorants in the family Pelecanus and the order Anseres. As Linnaeus defined it, this order encompassed a great diversity of waterbirds with both webbed feet and lobed toes. For the European shag, Linnaeus maintained *Graculus palmipes*, but for the great cormorant, he adopted *Carbo aquaticus*, which simply described the bird's black color and affinity for water. Two years later, the French naturalist Mathurin Jacques Brisson published *Ornithologie*, a six-volume set of meticulous ornithological descriptions and illustrations. It has so far had the final say on how cormorants are classified. Brisson grouped cormorants with anhingas, pelicans, tropicbirds, and boobies, and kept Pliny's *Phalacrocorax* to identify the genus to which these birds belong. Currently, at least eighteen genera have been described, but cormorants are still most typically classified in only one, *Phalacrocorax*. The European shag, now formally known as *Phalacrocorax aristotelis*, "Aristotle's bald raven," represents the long tradition behind the name.

Millennia have passed since the cormorant was first called a raven, and taxonomists recognized the difference between ravens and cormorants long ago. But the essential qualities that linked these birds in ancient Greece continue to link them today. The cormorant looks and sounds like a raven or a crow, a resemblance that lends it a special identity. Its somber black coat, combined with its intelligence when at work in the waters and seas, infuses it with a certain ominous aspect. Likewise, in its relations with humans, it is seen as a highly

Common raven

evocative figure of power and ill repute. These qualities remove the cormorant from the natural world inhabited by its waterbird relatives and transport it to the supernatural one in which the corvids dwell. Of the roughly ten thousand species of birds described thus far, few possess the power to achieve the cultural longevity attained by ravens, crows, and cormorants.[21]

With the advent of modern English, the cormorant's identification with the raven eroded, but its formal name continues to remind us of the bird's unique potency. It provides a particularly fitting bit of irony too. By bearing the description of an altogether different bird, and one that is bald to boot, the cormorant is misrepresented by even its scientific name, the point of which is to transcend language barriers and aid in universal recognition. But for numerous reasons, misrepresentation has become part of the bird's identity, and irony a constant companion. The cormorant's powerful but misleading scientific name is a perfect appellation for a bird that has remained enigmatic since the earliest efforts to describe it.

2

The Double-Crested Cormorant

An adult Double-crested Cormorant in full nuptial attire, viewed face to face at close range, is an uncanny and rather fearsome object.

—T. S. Roberts, *The Birds of Minnesota* (1932)

FROM A DISTANCE, the double-crested cormorant, *Phalacrocorax auritus*, is an unimposing figure, standing two and a half feet tall, with a wing span of approximately four feet, and weighing on average just under four and a half pounds. Its body, neck, and tail are all relatively long; the length of neck, especially, gives it a snakelike quality. Conversely, its legs, atop broad and powerful webbed feet, appear disproportionately short. Like most cormorants, it has iridescent black plumage that shines with an oily green, blue, or bronze gloss, depending on the light. The dark feathers of the wings and back are margined with a darker black, suggesting a pattern of scales and highlighting the bird's remarkably primitive appearance.

On closer inspection, a uniquely fierce aspect emerges. The naked skin surrounding the base of the bill and throat is a distinctive orange-yellow; the bill is mottled and possesses a pronounced hook at the tip. During the

Double-crested cormorant in a breeding display, eastern subspecies

breeding season, tufts of short feathers develop above the eye on both sides of the head, the feature for which this species has been named; the Latin *auritus* means "eared." These variably sized "double crests" are often difficult to see, but when visible, the crested fringe emphasizes the bird's wild edginess. The crests are the only seasonal change to occur in the plumage, but several dramatic changes occur in the bird's nonfeathered facial features. The inside of the mouth becomes a stunning cobalt blue, and the gular pouch (throat sac) becomes a more pronounced orange. The black iris of the eye turns a brilliant emerald or turquoise, and the pupil becomes encircled by a narrow silvery green. Completing the picture is the distinctive ringed eyelid, variably colored and often spotted with white, giving a jeweled or beaded appearance to the eye. At close range, this shock of blue, emerald, and orange against the iridescent black plumage has an arresting effect, and with the crests visible, the bird's formidable nature is striking.

When observed in its conspicuous spread-winged pose, common to several cormorant species, the cormorant acquires another potent aspect. In this notably bat- or vulture-like posture, the cormorant stands still and upright with both wings held out wide from the sides of its body. In this stance, frequently taken up after fishing, birds typically orient themselves toward the sun or the wind, presumably to dry their feathers or regulate heat loss and gain; some researchers have suggested that wing spreading occurs to heat up the bird's food and facilitate digestion.[1] Whatever the exact reason, the mysterious stance has an eerie, evocative quality, conjuring up images of crucifixions and vampires, and has fueled impressions about the bird's dark nature.

Observed on land, the cormorant offers yet another image. Moving with a waddling stride, and frequently making two-legged hops rather than walking, the cormorant is clumsy. Sometimes while clambering on rocks and trees, it holds on to whatever it can with its bill, increasing its awkward appearance. The cormorant's flight is not graceful either, sometimes being described as labored; but it is a fast flyer and can attain high cruising speeds. Its wing stroke is quick, like that of a duck, for which it is sometimes called the "crow-duck." Its wings are adapted for flapping, and it typically does not soar.

Underwater, however, there is nothing clumsy about it. In this environment the bird takes on its most potent and effective form. A streamlined swimmer, with wings held against its sides, the cormorant propels itself dolphin-like

through the water with simultaneous strokes of both feet. Its wettable plumage reduces its buoyancy, thus giving it the ability to exploit the water column both horizontally and vertically. Maximum diving depths are not known, but double-crested cormorants have been reported to dive in waters up to sixty-five feet deep, with a maximum dive duration lasting as long as seventy seconds.[2]

The Predator

A fast, agile predator, the double-crested cormorant is able to catch a myriad of aquatic animals. It occupies a place high in the food web, similar to top predatory fish such as walleye and pike. Representatives of more than sixty fish families have been reported as prey items, and while some sport and commercially valuable fish are consumed, the bulk of the diet commonly consists of forage or "trash" fish species, those not sought by humans.

The bird's power in the water and position in the food web is undoubtedly the most studied aspect of its natural history. As early as the 1830s, John James Audubon described the diet of cormorants he observed, reporting that "of the codlings especially they devour vast numbers." In general, fish about six inches long appear to be ideal for handling and are commonly taken, but almost any fish encountered that can be swallowed, will be; even those up to sixteen inches long occasionally make it on the menu.

Foraging habitats are likewise diverse; coastal bays, harbors, inlets, estuaries, rivers, interior lakes, ponds, reservoirs, and wetlands are all used. The most consistent features are shallow waters fairly close to shore; typically, such waters are less than twenty-six feet deep and within three miles of shore. They are also within eighteen miles of colonies and roost sites, but all these features can vary if a high-quality, consistently available food source is found. The double-crested cormorant easily adapts to locally abundant food sources and readily alters its foraging behavior to take advantage of fishing opportunities. Depending on how and where resources are encountered, birds will forage alone, in small groups, or in dense flocks. Bottom-dwelling and mid-water fish species (associated with inshore waters) as well as open-water, or pelagic, schooling fish species are all represented in the diet. But recent work in Saskatchewan to examine the position of cormorants in the food web found that the birds were not using the entire web, as would be expected for a truly generalist predator. Instead, birds from lakes had a diet that was consistently pelagic, comprised of species in the water column rather than from the lake bottom, suggesting that cormorants have a more restricted dietary niche in lakes than previously believed.[3]

The only rates of catch reported for this species are based on feeding trials with captive birds and on observations obtained at catfish ponds. In both circumstances, the extremely high catch rates reported show the birds' potential fishing efficiency. But these rates cannot be considered representative of their efficiency in natural environments, because artificial circumstances give cormorants unusual advantages. In particular, catfish ponds stocked with extremely high densities of fish in shallow water allow cormorants to meet their daily food requirements after about an hour of foraging. The kind of feeding activity observed in these simplified environments tends to reinforce the notion that the cormorant is a ravenous predator. For example, the researchers reporting the catch rate noted above commented that "the cormorants fed voraciously throughout their comparatively short feeding bouts."[4]

Despite this species' reputation for devouring fish stocks, its daily food consumption relative to its body mass is no greater than that reported for other fish-eating birds recognized as important fish predators. For instance, the daily food consumption of the double-crested cormorant is estimated to equal about 20–25 percent of its body mass, while that of common loons has been estimated at 26 percent and that of common mergansers at 31–39 percent.[5] Still, the double-crested cormorant is a particularly effective predator and is considered the ecological counterpart of the great cormorant, for which exceptional foraging efficiency is well documented. Frequently, when birds are shot and their stomachs examined, or when birds regurgitate after being disturbed in their colonies, large numbers of intact fish are observed, evidence suggestive of a rapid and successful fishing bout. The highly adapted body type of this species, combined with its ability to exploit a vast array of fish in diverse environments and to respond quickly to environmental change, probably results in relatively high catch rates in many locations. These exceptional attributes identify the double-crested cormorant as a potent human competitor, and have established it as the most significant avian predator on fish in North America.

The Engineer

Double-crested cormorants are powerful agents of change, engineers that design and transform their environment over time. Through their nesting and roosting activities, they exert a strong physical influence on the habitats they occupy. They are especially impressive nest builders, creating substantial nest structures both on the ground and in trees. In well-established ground-nesting colonies, nests can be maintained and added to over many years. The resulting nests are very sturdy and sometimes become tall, pillar-like structures reaching heights of six feet or more. In large colonies, these conspicuous nests can transform an otherwise barren substrate. Nests built in trees are also sturdy but far less substantial than the comparatively massive structures built on the ground. But the landscape changes that sometimes result from tree nesting can be far more dramatic. Nest trees often die as a result of cormorant activities, and over time a forested island can become a bare, scrubby one.

Changes to forested islands occur through a number of different mechanisms. Cormorants directly destroy vegetation by stripping leaves and breaking branches to build and line nests. In addition, branches break because of the birds' weight and movement in the trees. Less immediate but more profound changes result from the deposition of their acidic guano on the vegetation and soil under the nest trees. Leaves coated with guano can compromise a plant's ability to carry out photosynthesis, respiration, and evaporation, and changes in soil chemistry can be detrimental to tree health and understory vegetation.[6]

When cormorants occupy trees in dense numbers, they can affect both the abiotic components and the biotic community on islands in numerous ways. Losses in canopy cover and plant diversity may occur; the loss of native plant diversity may allow exotic, weedy, or invasive plant species to become established. If areas of bare earth result, they may be prone to erosion. As trees die and fall over, cormorants will remain on the island and begin nesting on the ground. But co-occurring waterbirds that nest primarily in trees, such as herons and egrets, are less flexible. These species may diminish in number and may eventually abandon the site altogether. Conversely, the newly created open substrate is suitable for ground-nesting species and frequently attracts gulls, terns, pelicans, and other birds.

These effects on vegetation are characteristic of all colonial waterbirds, but changes may be more obvious and extensive with particular species. What is unusual about cormorants is the extent and rapidity with which they can transform islands. To a large extent, this is a function of the cormorant's ability to form dense colonies and roosts quickly. Additionally, cormorants, and colonial waterbirds in general, typically select small islands for nesting, where mammalian predators are less likely to reside. On small vegetated islands, extensive and obvious changes from waterbird activity can occur rapidly.

In some areas, individual islands are highly valued by humans for their existing biological communities, and changes caused by cormorants are viewed as a threat to biodiversity and island ecology. Thus, the stage for human-bird conflict is set by another of the cormorant's unique attributes. The concern that cormorants may permanently eradicate or alter specific features of islands has become another significant factor leading to management of the species.

The Partner and the Parent

Those who have watched cormorants during the breeding season are always struck by the demonstrative actions and careful attention that paired birds give to each other, their nests, and their young. As pairs form, specific postures, stylized movements, and elaborate displays are used to advertise availability, attract mates, and claim territories. The bird's temporarily enhanced features, such as the cobalt blue mouth and striking emerald eye, are shown to their best advantage. But instead of a series of strange and bizarre behaviors, even casual human observers recognize much of their own nature while observing these birds. Early naturalists, less concerned about the danger of anthropomorphizing their subjects, frequently described the attention that courting and paired birds bestowed on each other as "amorous," "affectionate," and "joyous." In 1927, the wildlife biologist Harrison Flint Lewis, who did a great deal to increase understanding of cormorants, recorded the following observations about "an amorous couple" he watched in May:

> They were huddled side by side and . . . toying with one dead twig. . . . First one bird would take it in its beak and shake it much and then reach across the companion's back and place it in a position on the branches there. Then the other bird would take it . . . and repeat the performance, shaking the twig vigorously, as though it could hardly control its delightful feelings. This sort of thing went on for several minutes . . . with interludes in which the birds rubbed their bills, their necks, and the sides of their breasts together.

Later Lewis observed the presentation of a gift of seaweed, and saw one of the pair take a nap, "leaning affectionately against her swain's shoulder," while the other "stood his ground and looked about . . . as though proud of such

evidence of affection and confidence." So, too, Audubon and others noted the "joy" the birds seem to find in each other during "nuptial ceremonies." Audubon described sounds he heard during courtship "as if they came from a joyous multitude," and noted how, upon completion of displays performed on the water, the birds "swam joyously around each other, croaking all the while."[7]

More notable still is that this affectionate behavior not only occurs during courtship but lasts throughout the breeding season. Cormorants are monogamous, at least through a single season, and remain devoted to their mates. In 1936, the biologist Howard Mendall, in a major work on the natural history of double-crested cormorants, described the nature of the pair bond:

The petting or caressing that takes place when male and female are on the nest is one of the striking things to be observed. . . . This latter behavior consists of stroking of the mate's head . . . and gentle prods with the beak about the neck, back and wings. Rubbing of breasts between the paired birds is also of very frequent occurrence, and the whole performance is accompanied by low croaks and deep gurgling notes. . . . Such attention . . . is present as long as the birds remain together. Frequently I have seen mated Cormorants continue to spend several minutes at a time doing nothing but showing affection long after their young had left the nest.

Not even the welfare of the young, Mendall wrote, could "interfere with the interest that the parents have in each other. Polygamy is, I believe, unknown to them."

Cormorants approach the care of eggs and young with no less devotion. From the moment nests are begun until chicks fledge, both parents remain at the colony and take part in nest building and maintenance, and in incubating, brooding, and feeding young. In the early stages of incubation and brooding, one bird remains on the nest at all times; its mate, if not off fishing, frequently loafs along the shore or stands next to the parent on the nest. Typically, three to four eggs are laid, and then incubated for about twenty-eight days. During incubation, an adult rests eggs on the webs of its feet and carefully lowers its abdomen and breast on top of them, an incubation style unique to cormorants, shags, and possibly darters. Chicks, which hatch entirely without feathers, must be brooded continuously for the first two weeks. They are initially blind and feeble, barely able to move their limbs or lift their heads. During this stage, the feeding process is slow and deliberate; adults feed their chicks by regurgitation and take great care to transfer food gently to their delicate young.

As chicks grow, the number of feedings increases, reportedly peaking at nine to ten feedings a day, when chicks are sixteen to twenty-five days old. While confined to the nest, parents generally provide their offspring with meticulous care. On hot sunny days, they bring water in their gular pouches to the panting nestlings and often stand above them, sometimes with wings spread, to provide shade. I once came upon a nest in which an approximately three-week-old chick had obviously been trampled, the lower half of its body mashed into the nest. Though not long for this world, the parents had continued to care for and feed it.

The number of chicks a pair can raise varies widely. Values ranging from 1.5 to 3.5 young per nest have been recorded. As chicks become large, parents spend less time at the nest but continue to come back frequently to feed the growing young. At about four weeks old, chicks become mobile. If disturbed in ground nests, they will form crèches (large groups of young) and move about the colony, but they will continue to be fed by parents for several more weeks. Birds in tree nests remain in or near the nests until about six to eight weeks of age, when they can fly. Chicks become independent at about ten weeks. By then, the adults have made an enormous investment in their young, earning their reputation as superlative parents.

The Tendency to Aggregate

The highly social aspect of the double-crested cormorant's behavior is another feature that contributes to conflicts with humans. In each phase of the annual cycle, there is a strong tendency to aggregate in large numbers, which is believed to offer greater individual protection from predators. In winter, aggregations in the tens of thousands can develop on sandbars along the southern coasts; in the Delta region of Mississippi, huge concentrations occur near catfish farms in the isolated cypress swamps that birds use for night roosts. During migration, enormous flocks have been reported, ranging in size from tens of thousands to millions of birds.

Most spectacular, however, are the colonies that form during the breeding season. This species' amazing lack of need for personal space is then especially pronounced. Despite the small size of the islands frequently chosen for nesting, often only a half-dozen acres or less in area, colonies can consist of tens of thousands of breeding birds, along with untold numbers of subadults and nonbreeders. At a few sites, every square inch of available space appears to be in use. But even when ample space is available, nests are often built within a foot of one another when on the ground, or on the same branch when in a tree. Loafing birds, too, aggregate, standing in dense clusters along the shore. Occasionally, cormorants are even observed nesting on top of dead cormorants, an extreme act that speaks to the bird's incredible tolerance for close quarters, its drive to procreate, and its ever-opportunistic ability to appropriate a ready-made nest.

Over the season, numbers within the colony build. Birds continue to arrive over days or weeks, and the number of nests gradually reaches a peak. By midseason, if fish have been abundant and the colony has not been disturbed, many nests contain two or three chicks each, dramatically increasing the colony's overall density. As birds come and go throughout the day, their numbers fluctuate widely, but at any given time the sheer abundance of birds is impressive. Within the colony, apelike grunts, raucous guttural calls, gargles, hisses, whining, and begging provide constant background sound and further amplify the impression of numerical greatness. The shrill cries of gulls, common nest associates, add to the seeming chaos of the nonetheless thriving colony.

Large, well-established colonies are highly conspicuous. Sites used by ground-nesting birds are often distinctively whitewashed from years of accumulated guano, against which the masses of black cormorants and their dramatic nests stand out sharply. At sites where nesting has denuded or killed trees, cormorants are easily visible standing on bare branches or sitting on nests in the crook of a branch. Approaching such a colony from downwind, the reek of guano and rotting fish is detectable from quite a distance.

The Limits to Population Size: Ashmole's Halo and Other Mechanisms

Just how large cormorant colonies, and vertebrate populations in general, can become, and what places limits on their numbers, are questions that have long been of interest to biologists. Hypotheses revolve around the idea that population size becomes stable in relation to factors affecting population density. Density-dependent factors such as disease, predation, and competition for space or food become more important as the density of individuals in the population increases. One particularly compelling theory explaining seabird population sizes was proposed by the ornithologist Philip Ashmole in a seminal paper published in 1963. Ashmole's hypothesis grew out of an expedition to Ascension Island in the mid-Atlantic for the purpose of studying the biology of the numerous seabirds that were breeding there. Ashmole made several important observations about seabirds that led him to propose how their numbers might be regulated: adult seabirds suffered remarkably little mortality from predators, nesting space was not lacking, and no evidence indicated that disease was limiting population size. This left food as the remaining factor that could regulate numbers of birds.

Ashmole argued that food was most likely to operate as a limiting factor during the breeding season, when the fishing area is restricted. Because breeding birds can fly only so far and spend only so many hours away from their nests, foraging is concentrated around the colony. If the bird population is small, no changes will occur in the prey base and no competition for food will arise. But as seabird numbers increase, food supplies will diminish in the vicinity of the colony. This zone of depleted prey around a colony has been termed "Ashmole's halo."[8] When this condition occurs, competition among foraging birds may lower a colony's overall productivity by reducing brood size along with chick growth rates and survival; less efficient or less skilled birds may fledge fewer chicks or not breed at all. Ashmole suggested that these combined effects of competition regulated the numbers of most tropical oceanic seabirds and had the potential to regulate those of some species breeding at higher latitudes.

Ashmole's hypothesis is particularly relevant for double-crested cormorants. By relating cormorant population size directly to fish abundance, it identifies a concern that is at the heart of human-cormorant conflicts. And interestingly enough, the double-crested cormorant is the waterbird species that provides the strongest evidence for the existence of the halo. In the world of seabird research, this is significant because few empirical tests have demonstrated the existence of Ashmole's halo. Most evidence is indirect or circumstantial, deduced from observations that foraging distances or times become longer when the number of birds increases. But in 1985, the marine scientist V. L. Birt and colleagues conducted a study in the bays of Prince Edward Island in which scuba-equipped divers measured fish population density at various distances from two large double-crested cormorant colonies.[9] Fish densities were significantly lower in bays used by cormorants for fishing than in bays outside their foraging range. Although there was no evidence to indicate that prey densities would have been higher if cormorants had not been feeding in these areas, cormorants, as the major avian predator of bottom-dwelling fish in the area, were presumed to have lowered the densities where they were feeding.

Confirmation of Ashmole's hypothesis requires evidence that populations exhibit density-dependent growth, and such evidence involves years of historical data from breeding colonies. Relevant findings are limited for most species, but some evidence for density-dependent population regulation has been obtained for the intensely studied double-crested cormorant. An analysis of more than twenty years of data for colonies on Lake Huron by the fisheries scientist Mark Ridgway and his colleagues indicated that population growth stabilized between 2000 and 2003. The authors concluded that the population

had reached the biological carrying capacity of the environment and speculated that food was likely the factor limiting population growth. Additionally, a
decline in cormorant population density observed in 2003 was one of a number
of indicators, including significant changes in the abundance of several fish
species, pointing to an ecological regime shift in Lake Huron that occurred
during this time.[10]

A final aspect of Ashmole's hypothesis relative to cormorant population size warrants special notice. Observations of either the halo or density-
dependent population growth correlated with food availability convey the idea
that cormorants have depleted their prey, presumably by consuming them.
But some evidence suggests that foraging seabirds may deplete prey fish in
an area simply by disturbing them. In response to foraging by birds, fish employ escape responses that can include schooling, lateral movement (with fish
dispersing outward), or vertical movement (with fish swimming downward).
By dispersing prey resources, some of these responses can result in density-
dependent competition among cormorants that is not associated with increased
fish mortality.[11] This observation has important implications for interpreting
interactions between cormorants and fisheries, and provides a preview of the
complexity associated with assessing cormorant impacts to fish, which is discussed in chapter 12.

While the availability and abundance of fish are important determinants
in cormorant population size, predation and disease play a role, too, and may
operate on a population long before a zone of depleted prey develops. Adult
cormorants typically have few predators other than humans, although bald eagles will occasionally kill and eat adult birds. But predators can cause substantial losses of cormorant eggs and young, and in some areas where bald eagles
have experienced high rates of recovery, cormorants have become a significant
part of the eagle diet during the breeding season. At some of these locations,
predation has been substantial enough to limit reproductive success and, therefore, colony and population size. Intense eagle predation resulting in total reproductive failure has been observed in areas on the Pacific and Maine coasts,
at Lake Kabetogama in northern Minnesota, and on Rainy Lake, Ontario. On
the Pacific Coast, breeding sites safe from eagles appear limited, and eagle predation has been identified as a potential factor limiting population size in this
region.[12]

In areas where safe breeding sites do not appear to be limited, cormorants find a multitude of sites where they can successfully raise chicks. For

instance, in the Great Lakes, home to the largest freshwater island system in the world, cormorants have been documented to use only a few hundred of the more than thirty thousand islands in the system. In the continental interior, as cormorant populations become subject to intense management or illegal killing, birds have responded by colonizing new sites in the Great Lakes and on

Lake Winnipegosis, suggesting there is no shortage of breeding habitat in these regions. Similarly, in the wintering areas, roosting sites do not appear limited; new ones are colonized when birds are disturbed.[13]

Finally, cormorants are affected by numerous bacterial, viral, and parasitic infections. Newcastle disease, a contagious virus that affects domestic and wild birds, is the most prevalent one affecting this species. Sporadic outbreaks occurring in many portions of the range have significantly diminished numbers in local and regional areas. In some years, die offs of thousands of birds have been reported. Where such outbreaks occur fairly frequently, this disease can be a potent factor regulating population size.

Breeding Populations and Management Units

Because the range of the double-crested cormorant is so extensive, five geographically distinct breeding populations are recognized for management, conservation, and monitoring, but recent population changes have made their boundaries somewhat vague. In the western portion of the continent, birds breeding in southwestern and southern coastal Alaska and the Aleutian Islands make up the Alaska population, the smallest population unit, and do not come into significant conflict with humans. The most recent population data for this area, obtained between 1970 and 2000, estimated about 3,000 pairs. Moving south, birds breeding from British Columbia to Sinaloa, Mexico, are in the Western or Pacific Coast population. For the US portion of this unit, monitoring efforts by the Western Colonial Waterbird Survey extend from the coast inland as far as Montana, Wyoming, Colorado, and New Mexico. Data obtained mostly in 2009 for the United States and British Columbia estimated between 32,000 and 35,000 pairs, with 98 percent of the birds in the States. Mexico is not regularly or completely surveyed, but data obtained between 1968 and 1992 estimated approximately 14,500 pairs. Combined figures from Canada, the United States, and Mexico suggest the western population consists of approximately 46,500 to 49,500 pairs. Some conflicts over fisheries occur, particularly around the Columbia River estuary. But in other areas of this region, there is conservation concern about the status of the birds.[14]

Cormorants breeding from the Canadian Prairie Provinces and southern Ontario south across the Upper Mississippi Valley and the central Great Plains make up the vast Interior population. This is the most abundant—and from

N

- Breeding Locations 1970 to 2000
——— Continental Divide

1,000   500   0   1,000 Kilometers

Breeding distribution of the double-crested cormorant, 1970–2000 (Wires et al. 2001)

a human perspective, problematic—of the five breeding populations. Based on data compiled mostly between 2006 and 2012, the population is estimated at between 238,800 and 251,600 breeding pairs, with 67 percent of these birds occurring in Canadian waters. The Prairie Provinces and the Great Lakes support particularly large numbers. Across this region, cormorants are frequently observed fishing on both small and large lakes that are favored recreational and angling spots.

On the Atlantic Coast, double-crested cormorants breed from coastal Quebec and Newfoundland south through New England to Virginia. Data obtained mostly since 2006 provide an estimate of 93,615 breeding pairs, making this unit the second-largest breeding population. Approximately 70 percent of these birds occur in Canada, with large numbers in Quebec in the Gulf of St. Lawrence and the St. Lawrence estuary, and in the Maritime Provinces. Birds breeding here also come into conflicts with humans.

The fifth recognized breeding unit occurs in the Southeast along the Gulf and Atlantic Coasts and extends north to coastal North Carolina. In this region, breeding numbers are much smaller than those of northern populations. Data mostly from the late 1990s and later estimated this population at about 8,000 pairs, with about 84 percent of the birds breeding in Florida. But wintering populations in this region number in the hundreds of thousands; these birds are at the center of human-cormorant conflicts. The expansion of catfish and baitfish aquaculture in the southeastern United States, discussed in more detail in later chapters, is the ultimate example of how the overlap between the cormorant's distribution and foraging techniques, on the one hand, and human fishery practices, on the other, leads to inevitable human-cormorant conflicts.

Combining data from all five regions suggests that roughly 400,000 pairs breed in North America. In its fifth edition of global population estimates for waterbirds, the organization Wetlands International multiplied the number of breeding pairs it compiled by three to account for immature and nonbreeding birds, resulting in an estimated total population size of approximately 1,100,000–1,161,000 individuals, excluding those in Mexico. To provide some context for this figure, the population of the solitary breeding common loon is estimated at 607,000–635,000 adult individuals, while the colonial nesting ring-billed gull, one of the cormorant's most common nesting associates, is estimated at 2,550,000 individuals.[15]

Across the five geographic breeding regions, birds show differences in body size and coloration. Five races or subspecies have been described that correspond largely to distributional patterns. *Phalacrocorax auritus auritus* refers to the birds of the Interior and Atlantic Coast regions; they are moderately large in size and have curling black crests. *Phalacrocorax auritus floridanus* identifies the cormorant of Florida and the Southeast; these birds are small and dark with small black crests. The birds of southwestern and southern coastal Alaska and the Aleutian Islands are categorized as *Phalacrocorax auritus cincinnatus*; the largest in size, they have straight, mostly white crests. Birds breeding

Western subspecies showing white crests

on the Pacific Coast south of Alaska are designated *Phalacrocorax auritus albociliatus*, and are also fairly large and have partially white to all-white crests. Finally, *Phalacrocorax auritus heuretus* refers to birds that are resident on San Salvador Island and possibly other islands of the Bahamas, and on Cuba. These birds have the smallest and darkest form. Recent research by D. M. Mercer on the genetic structure of the continental population identified the Alaska cormorants as the most genetically distinct, including a potentially unique lineage restricted to the southwestern limit of the species range. Across the rest of the cormorant's range, the recognition of subspecies is not genetically supported. Instead, differences documented in this portion of the range appear to be due to geographic distance and isolation. Nevertheless, the five geographic regions and the use of subspecies names remain the central units by which populations are recognized, monitored, and managed.[16]

Among Other Birds

Cormorants seldom nest as a single species. Instead, they typically occupy a site only after it has first been colonized by other colonial waterbirds. In the vast Interior region, the colonial waterbirds that nest alongside cormorants include the distinctive aerial seabirds, the long-legged wading birds, and the elegant white pelican. In the first two groups, there are many representatives. Several species of ground-nesting gulls and terns, all resembling one another in form, size, color, and flight, share the rocky and sandy islands that cormorants often choose for nesting. At more forested sites, tree-nesting herons, night-herons, and egrets—also resembling one another in form, foraging style, and beauty—are frequent neighbors. The American white pelican has no similar-appearing family member, although the unrelated large white swans of the region provide some context. The occurrence of any of these birds in the landscape has never been questioned, and is perhaps made more acceptable by the fact that for any one of them there is another like it.

But the iridescent black cormorant is in a class by itself. It has no counterpart for color or form among this graceful company. It is further distinguished by its foraging style and nesting habits, being the only foot-propelled diver and the only colonial waterbird in the group that can nest equally well on the ground and in trees. The lack of any similar bird in the landscape adds to the cormorant's oddity and contributes to the sense that it somehow does not "belong." In fact, cormorants are often described as "invaders" in particular locations, even when they are native to a region, and are often considered "overabundant," even when they occur in relatively moderate or even small numbers.

The distinctiveness of the double-crested cormorant, its truly exceptional fishing and architectural abilities, and its tendency to concentrate in great numbers are features that have strongly influenced how the bird has been perceived by humans. Its unique appearance has inspired fear and disgust, and its fishing skill has inspired anger and hatred. Its transformative powers are judged destructive, and its populations have been deemed far too large. To understand the magnitude and power of these emotions and perceptions, consider that the number of cormorants legally destroyed between 1998 and 2011 is in the range of (or may even exceed) the total number of birds killed in some of the most catastrophic environmental events in recent history. For example, in the *Exxon Valdez* oil spill in Prince William Sound in 1989, estimates suggest somewhere between 250,000 to 580,000 seabirds died. By comparison, the number of double-crested cormorants killed in their breeding and wintering grounds under the two depredation orders and by permits since 1998 totals well over 500,000 birds, and immense yearly losses in reproduction have also occurred (see chapter 10). These losses suggest that human attitude remains the greatest threat this unique black bird faces today. So far, the cormorant's remarkable adaptability has allowed it to thrive despite the recent level of killing it has experienced, but it remains highly vulnerable to human persecution. The chapters ahead trace the rise and fall and the rise again of the double-crested cormorant, and the incredible effort humans have put into decimating populations of this unique species.

PART II

The Populations and the Perceptions, Then and Now

3

European Colonization and the Making of a Pariah

With the first streaks of dawn we beheld a sight that long will be remembered. From the hills there poured a steady stream of cormorants. . . . This stream poured from these hills continuously and reached as far as we could see, toward the bay of San Quentin. The stream was like a great black ribbon that waved in the breeze and reached to the horizon. It was truly a wonderful sight to see. The birds kept coming as though there were no limit to their numbers. . . . The flow of birds was continuous during the daylight hours. . . . The flow was unbroken—simply one steady stream going, all day, and a steady stream returning.

—Howard W. Wright, observations at San Martin Island, 1913

Eastern subspecies, *Phalacrocorax auritus auritus*

Unimaginable and Infinite Numbers

In 1604, Samuel de Champlain sailed along the southwest coast of Nova Scotia and visited several islands. One, west of Cape Sable, came to be called the Isle of Cormorants, "so named," the French explorer wrote, "because of the infinite number of these birds of whose eggs we took a barrel full." Of the nearby Seal Island group, Champlain wrote, "The abundance of birds of different kinds is so great that no one would believe it possible unless he had seen it—such as cormorants, ducks of three kinds, murres, wild geese, puffins, snipe, fish-hawks," and many others. Though Champlain did not identify the species of cormorant he observed, some of the birds were likely double-crested cormorants. More than two hundred years later, Audubon visited the Seal Islands and confirmed double-crested cormorants breeding there.[1]

Champlain's early seventeenth-century observations of the astonishing abundance of cormorants off the Nova Scotia coast were not exceptional. Rather, great numbers of cormorants thrived at many North American locations during the period of European colonization. In fact, several islands were named by European explorers for their abundance of cormorants. The earliest so-called Isle of Cormorants was in the Magdalen Islands area in the central part of the Gulf of St. Lawrence. The sixteenth-century English writer and explorer Richard Hakluyt visited the island in 1591 and recorded the earliest-known observation of cormorants in North America by a nonindigenous person. Fifteen years later, the French author and explorer Marc Lescarbot visited the northeast Atlantic region, "near Canso" (probably eastern Nova Scotia), and wrote, "The greatest abundance comes from certain islands where are such quantity of ducks, gannets, puffins, seagulls, cormorants and others that it is a wonderful thing to see [and] will seem to some almost incredible."[2]

In the eastern areas colonized by British settlers during the seventeenth century, several observations of the cormorant's great abundance were documented. In 1610, the winter of the infamous "Starving Time" at the first permanent English settlement, in Jamestown, Virginia, an anonymous writer noted that cormorants were among the birds in Virginia rivers that were "in such abundance as are not in all the world to be equaled."[3] In 1634, William Wood described how cormorants, which "bee as common as other fowles," were captured by Native Americans in New England. Three years later, Thomas Morton wrote of cormorants in New England, "There are greate store of Pilchers: at Michelmas, in many places, I have seene the Cormorants in length 3.0 miles feedinge upon the Sent" (*The New English Canaan*, 1637). In 1643, Roger Williams observed Native Americans returning with "vast numbers" of cormorants after a night's hunt in Massachusetts or Rhode Island, or possibly both.

In the eighteenth century, John Brickell published *The Natural History of North Carolina* (1737) and reported that cormorants "lay their eggs in . . . the islands, in the Sound and near the Sea Shoar in the Banks, and sometimes on high trees, as the Shags do," and were "as numerous all over these parts of North America as in any part of the World." In 1790 at the age of ten, Captain John Tanner was taken captive by the Shawnee, and in an account of his life he recorded how he killed "great numbers" of young cormorants at Lake of the Woods, Ontario, which, along with gull chicks, "covered" the island.

Such observations continued through much of the nineteenth century. In December 1820, Audubon noted of birds near Natchez, Mississippi: "We saw today probably *Millions* of those . . . Cormorants, flying Southwest—they flew in Single Lines for several Hours extremely high." In 1832, he reported from the Florida Keys: "Whole colonies of Cormorants had already built their nests, and were sitting on their eggs. There were many thousands of these birds, and every tree bore a greater or less number of their nests, some five or six, others as many as ten. The leaves, branches, and stems of the trees, were in a manner whitewashed with their dung." During his travels along the southwestern coast of Labrador and his visit to Cormorant Island, Audubon recorded that "the noise produced by the multitudes on the island was not merely disagreeable, but really shocking."

In 1878, Dr. Frank Langdon reported "boatloads" of cormorants being killed at the St. Marys Reservoir colony in Ohio. In the 1880s, John Macoun and Ernest Thompson Seton described the double-crested cormorant as an abundant breeder on Lake Winnipeg and Lake Winnipegosis. In 1887, Charles Whitehead visited a colony at the Minnesota-Iowa border, where observers reported "the air is jist black with em' an they're nestin on the island so yer can't see it for eggs." In 1891, George B. Sennett reported seeing a flock of cormorants in Minnesota that was four miles long and half a mile wide. The following year, Dr. P. L. Hatch published *The Birds of Minnesota*, which noted that cormorants bred in nearly all parts of the state, and were reported as "occasional to innumerable," depending on how close one was to breeding colonies.

Similar observations were made in the western portion of the continent. In the mid-1800s, thousands of double-crested cormorants nested in probably one of the largest Pacific Coast colonies at the South Farallon Islands, a foreboding collection of granite islands and rocky outcrops off the coast of Central California. In 1885, L. Turner reported double-crested cormorants were "abundant, resident and breeding in the Near Islands in the western Aleutians." In 1877, Charles Bendire reported "large numbers" breeding at Malheur Lake and on the Sylvies River in Oregon. In 1895, Corydon Chamberlin reported "an immense rookery [occupying] a place half a mile long" in Clear Lake, California. And at Magdalena Bay, Mexico, Walter E. Bryant reported that the numbers congregating were "almost incredible."

Although no continent-wide estimate is available for cormorants before 1900, their numbers impressed early explorers and naturalists, and the individual records suggest that the presettlement population may have encompassed millions of birds. But unlike some of the other abundant wild animal populations that were documented during European settlement, the multitude of cormorants during this time has received relatively little attention in accounts of America's natural history. This is partly because numbers of cormorants, while very large, appear less remarkable when compared to the astronomical numbers that were described for species such as the passenger pigeon, estimated in the billions at the time of Europeans' arrival, or the American bison, which roamed the continent in the tens of millions. For the latter creature, the sheer biomass that would have been present on the plains dwarfs populations of most species and is unfathomable today.

Where each of these species occurred in the landscape also affected impressions of their numbers. The passenger pigeon was a bird of the interior forest, living in colonies that stretched over hundreds of square miles, and migrating overland to and from wintering areas. The bison lived on grasslands in herds estimated to encompass hundreds of square miles and comprise millions of individuals. The cormorant, on the other hand, dwells in the vast wilderness of the waters and requires breeding and roosting sites safe from mammalian predators. It finds such refuge on small islands, places relatively tiny in land area and far less accessible to humans than the Great Plains or the forests. Although thousands of cormorants will jam themselves into a small area, island size and the number and distribution of fish in the water limit the number of cormorants at any one location. Additionally, sites used by cormorant colonies may be separated by many miles. Because cormorants dwelt on the fringe of human existence, the opportunity for seeing huge numbers of them was far more limited, but when experienced, no less remarkable.

Relentless Exploitation and a History of Carnage

The enormous concentrations of cormorants and other seabirds encountered by European explorers and colonizers were not just a cause for marvel. They were resources and commodities for unlimited harvest, a bonanza of good fortune for those seeking sustenance or profit. The breeding colonies, where seabirds gathered in great numbers with a multitude of flightless young, were especially profitable targets. For starving sailors and explorers making the long journey across the Atlantic, as for whalers, seal hunters, and fishermen, seabird colonies provided a source of fresh meat and abundant bait for fishing. In addition, they supplied oil for lamps, which was harvested from the fat of a variety of sea animals, including whales, birds, and fish. For all these uses, large numbers of seabirds were killed. During raids on colonies for bait, boatloads of birds were removed and entire colonies were cleaned out. Later, the trade in plumes and feathers became especially destructive, targeting a wide variety of species. Among seabirds, the feathers of gulls and terns were especially desirable. In fact, the demand for feathers during the nineteenth century was so great that a New York woman who negotiated with a Parisian millinery to deliver forty thousand or more bird skins hired gunners to kill as many terns as possible at ten cents a skin.[4] Egging, in which colonies were robbed of their eggs for human consumption, was another prevalent and lucrative practice. In many colonies, no nest was left untouched, and individual eggers sold eggs by the tens of thousands.[5]

Seabird colonies were often located on nearly inaccessible islands, but the lure of seabird products was worth the risk. In *Sea of Slaughter*, a five-hundred-year history of carnage along the northwestern Atlantic coast, Farley Mowat compiled records of the huge numbers of seabirds that could be killed in a single visit, sometimes in a matter of minutes. Some examples include the aforementioned record left by Lescarbot, who went on to state of his visit to the islands "near Canso": "We passed some of those islands where in a quarter of an hour we loaded our longboat with them. We had only to strike them down with staves until we were weary of striking." Later in the seventeenth century, the French explorer and colonizer Nicolas Denys raided the colonies at Sambro Island near Halifax, Nova Scotia, and recorded that his crew "found so great an abundance of all kinds [of seabirds] that all my crew and myself, having cut clubs for ourselves, killed so great a number . . . that we were unable to carry them away. And aside from these the number of those which were spared and which rose into the air, made a cloud so thick that the rays of the sun could scarcely penetrate it." In the early 1700s, the French fur trader and colonizer the Sieur de Courtemanche wrote about the north shore of the Gulf of St. Lawrence, reporting of the colonies there: "For a whole month they slaughter them with iron-tipped clubs in such quantity that it is an incredible thing." As guns became more available, the ease with which large numbers and entire colonies could be taken greatly increased, and the magnitude of seabirds harvested reached a whole new dimension.[6]

By the end of the nineteenth century, seabird populations were greatly diminished in many areas. Although most escaped annihilation, Mowat notes that populations survived along the northwestern Atlantic only "because of their astronomical original numbers, widespread distribution, or their ability to breed in remote or otherwise inaccessible places." The relentless exploitation that occurred has no better representative than the large and elegant great auk. Distributed in dense concentrations but on a limited number of islands, and completely unable to fly, the species was especially vulnerable to the Portuguese, English, French, and Spanish crews that fished, hunted, and traversed the North Atlantic. The great auk quickly became renowned for its meat, eggs, oil, feathers, and ease of harvesting, and on June 3, 1844, the last known pair was killed on Eldey Island, Iceland, although the seabird biologist and great auk expert Bill Montevecchi points out that "some lost and lonely surviving individuals wandered North Atlantic waters well after 1844." This annihilation represented the demise not only of a species, but also of the last flightless bird of the Northern Hemisphere.[7]

The "colossal destruction of animal life" described by Mowat during the period of European colonization encompassed a great diversity of marine and terrestrial animals. Seemingly infinite numbers of finned, furred, scaled, shelled, and tusked creatures were harvested until their populations vanished or became remnants of what they once were. Nor was this destruction limited to the northwestern Atlantic. At nearly the same time the great auk disappeared, a less well-known seabird was extirpated on the other side of the continent.

The spectacled cormorant, or Pallas's cormorant, the largest of all cormorant species, was a breeding bird with a limited distribution, endemic to the Commander Islands, west of the Aleutian Island chain in the Bering Sea. The only naturalist to see this bird alive was Georg Wilhelm Steller. Part of a crew exploring the area under the Danish captain Vitus Bering, Steller was aboard a ship that sailed to the area in 1741 and became stranded at what was later named Bering Island. The island offered access to a great abundance of wildlife, including sea otters, arctic foxes, and an apparently remnant population of sea cows, many of which had no experience with or fear of men. The crew, stranded on the island for nearly a year, used many of the animals for food, including the spectacled cormorant, which Steller described as "a special kind of large sea raven, with a callow white ring about the eyes and red skin about the beak." In midwinter, when other, more commonly eaten animals had become scarce or unavailable, the crew turned to these cormorants, one of which was large enough to feed three starving men.

Once off the island, news of its fantastic wildlife got out quickly, and the island was soon besieged by whalers, hunters, and fur traders. The sea cows, although too big to haul out of the water, were relatively easy to capture and became a favorite meat. They were harvested so thoroughly that only twenty-seven years after the island's discovery, the giant, peaceful sea cow was extinct. The island continued to be plundered, and in the early 1800s, the Russian-American Company imported Aleut natives to the island to harvest seals for the fur trade. The large spectacled cormorant, clumsy on land and possibly with reduced flying capabilities, was an easy catch. Valued for both meat and feathers, it was intensively hunted and by about 1850, the spectacled cormorant joined Steller's sea cow in the oblivion of extinction. The record left by Steller, six specimens, and two skeletons are the only evidence that this mysterious "sea raven" once fished the waters of the Bering Sea.[8]

The Persecution of Cormorants

During this period of unlimited exploitation, the widely and abundantly distributed double-crested cormorant was put to use for multiple purposes. Although notoriously fishy tasting, the cormorant was one of many birds that made it onto the dinner table in times of need. In addition, their flesh was highly sought for use as bait. Cod fishermen found cormorants especially attractive because their meat "hung together" well on hooks, and their chicks were easy to access and kill in large numbers. In addition, cormorant flesh was used as food for dogs and at fox farms. Colonies were plundered for valuable cormorant eggs, and although their feathers were not as valuable as those of terns, gulls, and egrets, cormorants were among the millions of birds killed in the late nineteenth century for the millinery trade. Even at breeding colonies where they were not the intended quarry of eggers, hunters, and feather traders, they nevertheless suffered dramatic losses. Typically more wary than other seabirds with which they nest, cormorants are usually the first to depart from and the last to return to their nests when human intrusions into colonies occur. Opportunistic gulls immediately consume the unattended cormorant eggs and small chicks, greatly reducing cormorant reproductive success.

While the double-crested cormorant was just one of many waterbirds facing multiple pressures during European settlement, toward the latter part of the era, its history began to diverge from that of other North American seabirds. For cormorants faced yet another threat that came to be synonymous with this bird: continual persecution for eating fish. Many seabirds were persecuted for this habit, but the extent to which cormorants were specifically targeted came to uniquely define this species. The cormorant family, in fact, epitomizes the persecution of fish-eating birds. In 1929, the first doctoral thesis on double-crested cormorants was written by Harrison Flint Lewis, providing an extraordinary benchmark in the species' history. Documenting the bird's distribution and abundance in the early twentieth century, Lewis's work is a monument to the effects of exploitation and deliberate persecution of cormorants. His goal was to describe the species' natural history and "to obtain such information as would cause its vivid avian personality to appear more clearly than before." Lewis focused on the most widely distributed and abundant subspecies, *Phalacrocorax auritus auritus*, the cormorant of the interior and eastern portions of North America. He examined and compiled records from Alberta to Quebec spanning the late 1500s to the early 1900s. Summarizing the cormorant's treatment by humans during this period, he wrote: "For one reason or another, or for no reason at all, Double-crested Cormorants have been continuously and severely persecuted. Adult birds have been shot, their eggs have been destroyed, their young have been massacred in flocks, their nests have been thrown into the water. . . . Almost every man's hand has been against them."

The late 1800s and early 1900s were a particularly deadly period for cormorants. Although seabirds were no longer considered important as bait for fishermen, Lewis and others report ongoing forays into cormorant colonies in which all eggs and chicks were smashed, clubbed, or otherwise destroyed, and as many adults as possible were shot. In *The Birds of Ohio*, William Dawson reported: "The Cormorant is becoming less and less common, even as a migrant, being fiercely persecuted by fishermen and thoughtlessly shot by every would-be sportsman who can hit a flying barn." Destruction was largely driven by the animosity of commercial fishermen, who perceived cormorants as their nemesis. This belief was fueled in part when commercial fishermen began using pound nets in the 1830s. Highly effective for entrapping large concentrations of live fish, these nets were widely used in the nineteenth and twentieth centuries. Trapped fish are quickly discovered by fish-eating birds, and fishermen frequently observed cormorants taking fish from the nets. In addition to those losses, there were massive declines in fish populations in many parts of the continent due to the enormous harvests by European and American fishing fleets. Competition from cormorants at the pound nets led fishermen to launch a deliberate, nearly systematic effort to extirpate the birds. Lewis described the bird's history during this time as "a history of persecution and of the gradual abandonment of one breeding place after another."

In the first quarter of the twentieth century, large numbers of breeding double-crested cormorants could still be seen at a few locations that experienced less intense European settlement. The American ornithologist Arthur Cleveland Bent traveled to the remote northern portion of the cormorant's range in 1913, where he reported a colony of 1,500–2,000 nests on Lake Winnipegosis, Manitoba, the largest cormorant colony he had ever seen. During the same year, the ornithologist Howard Winslow Wright visited the west coast of Mexico, another area where relatively little human influence had as yet been exerted, and discovered the largest cormorant colony ever documented. On Isla San Martin, off the west coast of Baja California, Wright estimated an astonishing 348,840 nests. The ornithologist J. R. Jehl later revised the estimate down to 213,500 pairs, but by either measure, the colony that Wright observed was clearly enormous, dwarfing any known in more recent times and providing an inkling of a now-unimaginable past for this species.[9]

Records such as Wright's mark the end of an era for cormorants, and by the time Lewis was writing, they had vanished from or were greatly diminished in many places. Lewis's thesis included a list of locations where breeding cormorants could still be found and from where they had disappeared. In the northwestern Atlantic, they were gone from multiple sites in the Gulf of St. Lawrence, the southwestern coasts of New Brunswick and Nova Scotia, and coastal Maine and Massachusetts. Likewise, they disappeared from numerous locations in all three Prairie Provinces, as well as from sites in Iowa, Ohio, Illinois, Minnesota, the Dakotas, Wisconsin, and Arkansas. He estimated a total breeding population of 26,586 birds in interior and eastern North America, and although he acknowledged that his estimate was incomplete, particularly in the northwestern portion of the range, he nevertheless concluded that cormorants were far less abundant in this portion of North America in the late 1920s than they had been formerly. A review of records for the region that were unavailable to Lewis indicates declines had occurred in other locations as well. At Lake of the Woods, Ontario, where Tanner had observed "great numbers" in the late 1700s, numbers had been diminished to fifty breeding birds by the time Lewis was writing.

Seven years after Lewis published his dissertation, Howard Mendall completed a second major study on double-crested cormorants. Focused on the species' history and status in New England, it included multiple records documenting the cormorant's breeding distribution before 1900. Mendall identified the breeding range as extending from Maine to possibly as far south as

Rhode Island; more recently obtained archeological records confirm that cormorants bred as far south as Boston Harbor in about 1500.[10] Mendall included William Wood's early seventeenth-century record of cormorants in the area, which indicates that animosity toward these birds, although particularly prevalent in the late 1800s and early 1900s, was a centuries-old sentiment: "Cormorants," wrote Wood, "destroy abundance of small fish, these are not worth the shooting because they are the worst of fowles for meate, tasting rank and fishy; [but] No ducking ponds can affoard more delight than a lame Cormorant and two or three lusty Dogges."

As in the interior region, cormorants were heavily exploited and persecuted across New England. At breeding colonies, they were subjected to intensive shooting and egging, and were indirectly impacted by the intense harvesting of gulls and terns for food and feathers. Cormorants were also persecuted away from colonies at locations where they gathered in large numbers. In 1834, the English ornithologist Thomas Nuttall reported that in Massachusetts Bay, "at the approach of winter they are seen to assemble in numerous dense flocks, so that several dozen have been killed at a shot." By the late 1800s, the cormorant, along with eight other seabird species, had been extirpated as a breeding bird from New England.

Exactly when cormorants disappeared there is uncertain. In 1883, Everett Smith reported that none bred along the coast of Maine. In 1884, Spencer Fullerton Baird, Thomas M. Brewer, and Robert Ridgway published *The Water Birds of North America*, and stated that the cormorant was "believed once to have been resident on the northern shores of Massachusetts, but long since to have been driven away." Then, between 1892 and 1896, a small number of cormorants, fewer than ten, were observed nesting at Black Horse Ledge, Maine. During this period, egg collectors visited the site and robbed the few cormorants of their eggs each year the colony was active. In 1896, just two nests were found: one was incomplete, and collectors had robbed the other of its eggs. This was the last nesting record obtained in New England for decades to come.[11]

In eastern Canada, substantial declines were obvious along the coast of Nova Scotia, where cormorants had been so abundant during Champlain's visit that islands were named for them. By the mid-1920s Lewis reported only sixty-seven pairs at two sites. Likewise in Quebec, cormorants were extirpated from many locations, mainly by human persecution and exploitation. Lewis noted that cormorants "supplied fresh meat for Indians, fresh eggs for local residents, and a considerable quantity of food for sled-dogs and captive foxes . . . and that some colonies lose practically all their young every year." Although information is limited, declines appear to have occurred in Newfoundland and New Brunswick as well.

A review of the cormorant's history in the Pacific Coast region was prepared in 1995 by the biologist Harry Carter and his colleagues; it revealed a similar pattern of population change and loss of breeding colonies on both the coast and in the interior. In Alaska, the large-scale introduction of arctic and red foxes for fur farming, which peaked in the early 1900s, is believed to have led to significant declines in the numbers of many seabird species, including double-crested cormorants. Further south along the coast, shooting for sport, persecution by fishermen, intensive egging, and exposure to oil from tankers flushing their tanks near islands used by seabirds resulted in substantial declines in cormorant numbers and colony sites. Changes that occurred at the Farallon Islands are particularly well documented, providing one of the best records of the numerous pressures cormorants faced in that area. South Farallon, the largest and most accessible of the islands, was visited by William Dawson in 1911, by which time the number of double-crested cormorants had declined by the thousands—only thirty-five pairs remained. Dawson remarked, "The story of the steady persecution to which the confiding members of this historic colony have been subjected would not make a pretty one in print." Mexico appears to be the one exception to trends observed elsewhere, and cormorants remained abundant in some areas. Besides Wright's record of the enormous colony at Isla San Martin, the zoologist Joseph Grinnell, in a description of the birds of Baja California published in 1928, reported that double-crested cormorants were "resident abundantly along the whole western side of the peninsula."[12]

While declines on the Pacific Coast probably resulted in the complete loss of colonies in regions first settled by nonnative peoples (for example, southern British Colombia, Puget Sound, the Columbia River, San Francisco Bay), in some areas the arrival of Europeans may have increased seabird numbers. Before European colonization, accessible seabird colonies were likely visited regularly by native people to obtain birds and eggs. Radiocarbon dating of material from island and mainland locations indicates that native peoples gathered food from coastal rocks and islands for thousands of years. Therefore, many seabirds may have existed in much lower numbers at more accessible sites when Native Americans were present. With the arrival of European settlers and the elimination of native peoples, seabirds may have been able to slowly expand or reoccupy former breeding areas. For example, double-crested cormorants were first recorded breeding on the San Juan Islands in the inner coastal waters of Washington in the twentieth century. But a recent examination of archeological evidence from Watmough Bay in the San Juan Islands by the zooarcheologist Kristine Bovy, published in 2011, provides ample evidence of young birds, indicating that a breeding colony was present some fifteen hundred years ago. The colony's disappearance sometime later may have been due to pressure from aboriginal people rather than European settlers.

Although a variety of human actions led to the cormorant's demise during this period, the role of persecution specifically aimed at extirpating these birds stands out as the most remarkable factor involved. An especially poignant commentary on this aspect of the cormorant's history was included in a report on cormorants in Minnesota and North Dakota prepared by F. M. Uhler in 1925 for the Bureau of Biological Survey. At the time, only one cormorant colony remained in North Dakota, in the northeastern portion of the state at Stump Lake. The lake was drying up and becoming increasingly alkaline, a condition that had killed all the fish except sticklebacks. Uhler investigated the cormorant's diet and concluded that cormorants were surviving mostly on salamanders, but were nevertheless persecuted for the loss of the freshwater fish. Uhler commented:

> From a humanitarian perspective, considering the fact that this is apparently the only rookery left in North Dakota, and also that the birds have nested here since long before the arrival of the white man, it seems difficult to understand how anyone with the semblance of a sense of justice can begrudge a few fish, even of the choicest variety to any species of bird not obviously detrimental to the interests of man. Should we not, in some small degree, apply the same "right of property" by which man seems to so jealously govern himself to the feathered inhabitants of the earth? In the opinion of the writer, the ancient savage instinct toward wanton destruction certainly has no place in modern civilization.

In identifying this "ancient savage instinct" Uhler highlights the definitive theme in the cormorant's recent history, not just in North Dakota but across North America.[13]

The Making of a Pariah

Put a knife to your throat if you are given to gluttony.
—Proverbs 23:2

Of the myriad seabird species European explorers encountered when they arrived in North America and settled across its lands, why did the double-crested cormorant become the focus of such hostility? Although cormorants eat fish, and plenty of them, so do other seabirds. What was it about the cormorant in particular that inspired raids on colony after colony for no other reason than to rid lakes, rivers, and seas of this feathered presence? One factor is indeed the bird's ability to catch fish rapidly. Evidence that European settlers considered cormorants destructive birds in this regard is apparent in the comments of some early chroniclers about the cormorant's diet and their impressions of its nature. William Wood's observation from 1634 that "cormorants . . . destroy abundance of small fish" reveals that the cormorant was judged and found guilty early in the period of European settlement. Likewise, John Brickell's notes on cormorants in 1737—"they catch vast quantities of Fish, which is their only Food, and whereof they are very ravenous and greedy"—identifies the perception of the cormorant as a bird with an excessive and insatiable appetite and a necessarily selfish character. Nearly a hundred years later, in describing the Florida cormorant, Audubon stated that "their appetite is scarcely satiable, and they gorge themselves to the utmost at every convenient opportunity."

Ample evidence exists to indicate that early explorers and settlers arrived on the continent with strong feelings about this family of birds. These attitudes stemmed largely from long-held perceptions about the bird's appetite. As early as the fourteenth century, Geoffrey Chaucer, in *Parliament of Fowls*, referred to "the hote cormoraunt of glotonye," paraphrased as "the greedy, gluttonous cormorant" or "the cormorant, hot and full of gluttony." By 1531, the cormorant's association with gluttony was apparently so well established that the *Oxford English Dictionary* cites the use of *cormorant* figuratively to indicate "an insatiably greedy or rapacious person." By the end of the sixteenth century, the cormorant's gluttony was immortalized through William Shakespeare's references to its voracious, destructive appetite. In *Richard II*, Richard is compared to an "insatiate cormorant." In *Love's Labour's Lost*, "Time" is described with the adjective "cormorant devouring." In *Troilus and Cressida*, the Trojan War is described as a "cormorant war" in which all things dear are "consumed in hot digestion." Finally, in *Coriolanus*, Shakespeare retells Aesop's "Fable of the Belly," in which "all the body's members / Rebell'd against the belly" because of its unwillingness to share. The belly is described as a "cormorant belly," emphasizing its truly selfish and gluttonous nature. About the same time these plays were written, cormorants were formally declared a pest of the crown by Queen Elizabeth I, and in 1603, the last year of her life, she placed a bounty on them. The image of the cormorant as a glutton persisted, and some two hundred years later this well-known aspect of the bird was used by another literary great, Samuel Taylor Coleridge, who described himself as "a library cormorant" in regards to his having read almost everything.

The perception of the cormorant as a glutton has far-reaching implications. From the Latin *gluttire*, "gulp down" or "swallow," gluttony by itself refers to the overconsumption of food. But in the Christian tradition, this lack of self-control not only indicates overindulgence but also suggests a loss of reason and moral virtue, a serious enough failing to be counted as one of the seven deadly sins. Thus, as the embodiment of gluttony, the cormorant was derided not only because of its perceived overindulgence in food, but also because such behavior has a dark, unholy aspect.

Within the context of the massive slaughter and waste of wildlife throughout North America during the period of European colonization, particularly during the nineteenth century, this perception also nods to the cormorant's ever-present companion, irony. Destroying an unfathomable number of creatures, including birds, bison, and sea cows, in just a few decades, not to mention millions of native peoples, the voracious new Americans readily found the cormorant a convenient scapegoat for their own sins of greed and gluttony.

> Gloomy, voracious, filthy.
>
> —"The Bible Birds" (1867)

Something besides perceptions of gluttony has been involved in the cormorant's becoming a pariah. Traveling further back in time, a long history of the bird's association with darkness unfolds. As we have already seen, its name and identity for most of the last two thousand years were derived from its similarity to another bird with pronounced ties to darkness, the raven. But even earlier traditions had identified the cormorant as malevolent. Early books of the Bible, written some three thousand years ago, describe cormorants as unwholesome and hateful. In Leviticus, the Lord identifies clean and unclean food to Moses and Aaron, telling them which birds they may and may not eat, and the cormorant is among the birds that they "are to detest and not eat because they are detestable" (11:13–19). Similarly, cormorants are identified in Deuteronomy as unclean and unfit as food (14:17). In chapter 34 of Isaiah, which contains prophecies of doom, the land will be inhabited by the unclean animals of the wilderness, of which the cormorant is one. In Zephaniah, the inheritance of the earth by wild beasts is reiterated, and the cormorant is connected more directly with sin and evil: "And flocks shall lie down in the middle of her [Judah], all the beasts of the nations: both the cormorant and the bittern shall lodge in the upper lintels of it; their voice shall sing in the windows; desolation shall be in the thresholds" (2:14).

In Revelation, the role of birds in the "demise of man" is described: "And I saw an angel standing in the sun; and he cried with a loud voice, saying to all the fowls that fly in the midst of heaven, Come and gather yourselves together unto the supper of the great God; That ye may eat the flesh of kings, and the flesh of captains, and the flesh of mighty men, and the flesh of horses, and of them that sit on them, and the flesh of all men, both free and bond, both small and great" (19:17–22). Thirteenth-century artists in England and France, working in a time of great religious and political conflict, illustrated this passage by transforming St. Francis of Assisi's famous "Sermon to the Birds" into apocalyptic representations. Several of these depict cormorants or cormorant-like birds diving and feasting on the flesh of men while an angel stands by and St. Francis gives his sermon.[14] Designated as unclean and detestable and identified in early prophecies of doom, the cormorant is, unsurprisingly, represented as one of the birds of the Apocalypse.

Why the cormorant was originally deemed an unclean animal or chosen as a harbinger of doom is not clear. The bird's early connection with sin may have resulted from its "gluttonous" behavior or may have arisen simply from the bird's powerful aspect. Whatever the reason, given the biblical tradition and its legendary appetite, the prehistoric-looking cormorant became a logical choice for representing Satan himself. In his seventeenth-century epic poem *Paradise Lost*, John Milton does exactly that. Satan, as he journeyed to Paradise, "flew; and on the Tree of Life / the middle tree and the highest there that grew / Sat like a cormorant . . . / . . . sat devising death / To them who lived." The combination of special qualities ascribed to the cormorant, such as insatiability, greed, desolation, and detestment by God, along with its snakelike appearance, makes it a natural embodiment of evil. It appears again as a bird of ill fortune, "Portending ruin to each baleful rite," in William Wordsworth's *Ecclesiastical Sonnets* (1822). In a note to that work, Wordsworth comments: "The cormorant was a bird of bad omen."

In North America, these attitudes quickly became attached to the species found there. In fact, recognition of the cormorant's evil character appears to have been an important enough teaching for it to be described in the *Ladies' Repository*, a monthly periodical devoted to literature, the arts, and the doctrines of Methodism. In 1867, the magazine published "The Bible Birds," which described the cormorant as "gloomy, voracious, filthy; famous for plunder, and very fond of fish, which it devours with surprising gluttony. It has a rank and disgusting smell, and is said to be, even in its living, healthful state, more fetid and offensive than carrion. In Isaiah and Zephaniah, this foul bird is spoken of as inhabiting the ruins, in connection with the owl and the bittern."[15]

The Blackness of the Bird

In 1903, the Dutch ornithologist J. P. Thijsse identified the great cormorant as "literally and figuratively the *bête noire* of fishermen," and so highlighted yet another significant theme in the cormorant's demonization.[16] An essential attribute of the bird itself—its glossy black plumage—rendered it a target of irrational human sentiment in Europe and North America. In many cultures, black animals in general have special symbolic associations. In medieval Europe, a black coat was especially meaningful, signifying an obvious and potent sign of difference and otherness. Animals so colored were frequently associated with evil, witchcraft, magic, superstition, premonition, otherworldliness, and all things dead. The blackness of certain animals in particular, such as cats, sheep, and birds, was often considered a mark of the devil. The presumed alliance with the dark world implied by black fur or feathers had serious consequences, and resulted in the deaths of inestimable numbers of animals. A potent example is the persecution of cats that occurred in medieval Europe. In 1233, Pope Gregory IX issued a papal bull in which he claimed the devil had appeared in the form of a black cat. This was the first official church document recording this association, and arguably part of the justification for the intense persecution of cats that ensued for hundreds of years.[17]

For its creation as a pariah, the importance of the cormorant's black coloration has its greatest resonance in the most obvious instance of persecution due to color: human discrimination and slavery based on racial difference. Persecution of blacks by whites resulted from the most fundamental distinction—"otherness" inherent in skin color. The symbolism associated with light and dark, and good and evil, historically played a large role in how light-skinned or "white" Europeans treated, and justified enslaving, darker-skinned people from other cultures. In this context, it is of no small significance that by the 1920s, among the many derogatory names the cormorant came to be called, that of "nigger goose" was popular in the southern United States, and was also in use in northern states as late as the 1950s. In a paper published in the *Passenger Pigeon* in 1951, the naturalist George Knudsen described visiting a Wisconsin rookery where the baby cormorants "covered in a very wooly black down" reminded him of "pickaninnies." He went on to comment about the persecution of cormorants and great blue herons because of their fish-eating habits: "It seems that everyone is anxious to exterminate the 'nigger goose' and the 'shitepoke' [great blue heron]." In literature from the western part of the continent, no references to this name have been found, but a story is told in which the association was apparent. During a visit to Lower Klamath Lake, California, in 1915, a field team approached a group of young cormorants; despite being spoken to "in the Negro tongue . . . the members of the colony waddled off suspiciously." The use of the *n*-word—arguably the most infamous term in the English language—as a modifier for the cormorant reveals the importance of the bird's color in its persecution more definitively than could an entire book on the subject. Invoking a centuries-old tradition of discrimination and terrorism against those that were "other," it identifies the special kind of hatred that leads to lynchings, pogroms, and genocide, and provides a powerful context in which human-cormorant conflicts must be considered.[18]

4

From Audubon to Conservation

The First Wave of Recovery

ON APRIL 26, 1832, John James Audubon celebrated his forty-seventh birthday by painting the double-crested cormorant in the Florida Keys. Believing cormorants in the region to be distinct from those elsewhere in the country, Audubon identified the bird as the Florida cormorant. In his portrait, a single cormorant looms in the foreground, perched on a craggy bit of dead wood rising out of unsettled water. In the distant background, patches of mangrove forest surround the perimeter, suggesting that the cormorant is in a bay or an inlet that connects to the open sea. More cormorants loaf on tiny islands or bits of land jutting out of the water, and upon close inspection, many cormorants are visible in the mangrove trees. The bird in the foreground is captured in motion, its sinewy neck arched and reaching, and its bill is wide open, as though calling out. The eye is prominent and intense, and the bill's sharp hook is pronounced. The large black feet clutch the small bit of wood on which the bird perches, and their webbed edges hang above the water. The wing is draped along the bird's side, and the feathers end in sharp, jagged edges, a quality likewise reflected in the feathers of the tail and the tiny crest above the eye.

The text accompanying the illustration, like most of the "ornithological biographies" Audubon wrote for the birds he painted, recounts the dramatic "episode" his observations entailed. A careful description of his subject's environment, habits, behavior, and anatomy reveal an intelligent and interesting bird. Both the portrait and the text capture the wild vitality Audubon encountered. The text, however, goes beyond the watercolor to elaborate the powerful effect the environment and the bird had on the artist:

> The Keys were separated by narrow and tortuous channels, from
> the surface of the clear waters of which were reflected the dark

mangroves, on the branches of which large colonies of Cormorants had already built their nests, and were sitting on their eggs.
There were many thousands of these birds, and each tree bore a
greater or less number of their nests, some five or six, others perhaps as many as ten. The leaves, branches, and stems of the trees,
were in a manner white-washed with their dung. The temperature
in the shade was about 90 degrees Fahr., and the effluvia which im-
pregnated the air of the channels was extremely disagreeable. Still
the mangroves were in full bloom, and the Cormorants in perfect
vigour. Our boat being secured, the people scrambled through the
bushes, in search of the eggs. Many of the birds dropped into the
water, dived, and came up at a safe distance; others in large groups
flew away affrighted; while a great number stood on their nests and
the branches, as if gazing upon beings strange to them. But alas!
they soon became too well acquainted with us, for the discharges
from our guns committed frightful havoc among them. The dead
were seen floating on the water, the crippled making towards the
open sea, which here extended to the very Keys on which we were,
while groups of a hundred or more swam about a little beyond
reach of our shot, awaiting the event, and the air was filled with
those whose anxiety to return to their eggs kept them hovering
over us in silence. In a short time the bottom of our boat was covered with the slain, several hats and caps were filled with eggs; and
we may now intermit the work of destruction. You must try to
excuse these murders, which in truth might not have been nearly
so numerous, had I not thought of you quite as often while on the
Florida Keys, with a burning sun over my head, and my body ooz-
ing at every pore.[1]

This account, directly addressing the reader, is particularly interesting because
of its self-reflective and intimate nature. For Audubon, killing large numbers
of birds for a day's work was by no means exceptional. Rather, most of the
stunning watercolors presented in *Birds of America*, a collection of dramatic,
life-sized portraits representing one of the greatest ornithological achieve-
ments of all time, were painted from freshly dead birds that Audubon killed or
hired others to kill. Early in his quest "to paint every bird in the United States
and its territories," he pioneered an innovative technique for creating his dramatic illustrations. He would first shoot birds with fine shot so as not to damage

their plumage, and then use wires and threads to hold their bodies in lifelike poses, allowing him to animate his subjects in three dimensions. This technique immediately distinguished his paintings from the wooden poses and flat presentations of other artists, but often required killing dozens of specimens to complete one portrait. Within this context, it is all the more noteworthy that Audubon should seek to be excused for his self-described "work of destruction" and to acknowledge that, even given the requirements of his painting technique, his methods had been excessive that day in the Keys.

More interesting still is that this overzealous behavior should occur in a cormorant colony. According to the artist, his behavior was provoked by the conditions of the environment rather than by the birds themselves. Nevertheless, the rest of the account leads to a sense that Audubon felt a mix of appreciation and disgust for these birds. To describe their "nuptial ceremonies," he used words like *joyous* and *graceful*. But he also commented that the cormorant's inability to charm with a song, offering only grunts instead, along with its "rank and fishy" taste and its tendency to "eat and mute [defecate] inordinately," conspire to make these birds "less pleasing objects than other feathered pets." In this way, of course, he reflected the popular attitude toward cormorants.

In June 1833, Audubon painted a second portrait of the bird and wrote another interesting account describing it. This depiction was based on cormorants he observed along the southwestern coast of Labrador, which, unknown to him, belonged to the same species he had already painted in Florida. In the later portrait, a single cormorant stands on a bare rock and looks out over the sea. The bird is drawn in sharp contrast against an immense sky and appears to be contemplating the view before it. Everything about this cormorant is softer than the earlier portrait. The bird's stance is more upright, the eye is less intense, the feathers lack jagged edges, and the neck has no snakelike suggestion. Far in the background, cormorants loaf on a sandy island, and a few fly low over the water. The lone cormorant in the foreground is regal; set against the vast space of sea and sky, it perfectly captures the isolation and austerity of the place to which it belongs. So, too, does Audubon's powerful description:

> The objects that more especially attract the notice of the voyager . . . are the numerous low islands covered with countless multitudes of birds, that have assembled there for the purpose of reproduction. Some miles farther you see a ridge of craggy and desolate cliffs, emerging from the sea, and presenting the appear-

ance of a huge granite wall. . . . From the hoary summit of this bulwark the view is grand beyond description. . . . In that land, man may for weeks, even months, seek for his kind in vain. The deep silence that reigns around him during a calm, seldom fails to bring sadness to his heart, as his eye grows dim with gazing on the wilderness. Should the northern gale issue from its snowy chambers, darkness follows in its train, and should its whole fury pour upon you, melancholy indeed but be your lot.

To the low islands above alluded to, the beautiful Cormorant represented in the plate before you, resorts each spring, for the purpose of breeding.[2]

In this account, Audubon appears to have decided that despite the attributes that made it "less pleasing than other feathered pets," the cormorant's beauty nevertheless made it a pleasing bird to behold. This reevaluation is all the more profound because it suggests that Audubon transcended his former revulsion, seeing it as something distinct from the bird's intrinsic beauty and value. Like the first account, the second identifies some of the more disagreeable aspects of entering a colony and handling the birds. It describes the chaos that ensued in the colony when his crew landed on the island, and the "mass of putridity" he encountered on the surface of the rock where the birds had likely been nesting for centuries. He notes how, when approached, the young birds "twisted their necks in the most curious manner, reminding one of the writhing of a snake, and when seized they muted so profusely as to inspire disgust." Nevertheless, the word *beautiful* is used three times in his description, and Audubon closes his account by referring to his presentation of "the figure of a beautiful male in its perfect spring plumage."

Side by side, the two portraits and their accompanying texts are striking and effective. They bring to life an interesting and attractive bird and convey a sense of the rugged wilderness in which cormorants thrive. At the end of the later account, Audubon notes that his portrait "is probably the only representation of the bird yet presented to the public, and the same remark applies to the Florida Cormorant." Looking more broadly at the impact of Audubon's work, the same can be said for several species of birds—for many, Audubon's paintings provided the first images of them that the public had ever seen. And by fully embodying his subjects, Audubon's unique portraits provided far more than just a glimpse at unusual birds. His paintings were "personalities in motion," and they immediately captured the attention of those who saw them. His writings, too, contributed important details of natural history, and his evocative style transported readers to otherworldly places inhabited by fantastic creatures. Audubon's attention to the natural beauty of birds and the places they lived brought them to life as no previous works had done. While Audubon killed thousands of birds simply to paint and describe them, his portraits, in a stunning bit of irony, planted the seeds of awareness that fostered an appreciation for birds in and of themselves.

By the late 1830s, most of Audubon's paintings were complete, and with the publication of *Birds of America* in 1838 he became something of an overnight success. To make the work more accessible to a larger audience, Audubon published a royal octavo (10 in. by 6.25 in.) edition in 1844, which was much smaller and less expensive than the original, "double elephant folio" (39.5 in. by 28.5 in.) version. In the introduction to the first edition, he wrote that it was his desire to make *Birds of America* available at "such a price, as would enable every student or lover of nature to place it in his library." The beauty of the portraits and the power of the accompanying descriptions helped popularize birds, and public interest in watching and studying them greatly increased from this time onward.

Development of Protection

At about the same time this appreciation for birds was developing, other important influences on the perception and appreciation of nature in North America were also evolving. Shortly after Audubon's royal octavo edition was published, Henry David Thoreau gave his lecture "Walking," presenting the idea that "in Wildness is the preservation of the World." In 1872, the establishment of Yellowstone National Park represented the world's first instance of large-scale wilderness preservation. Despite these developments, however, decades passed and incredible numbers of wild animals were killed before there were any remotely serious efforts to protect birds. By the late 1800s, an estimated five million birds a year were being killed for the millinery trade alone, the almost inconceivably enormous population of passenger pigeons had been laid to waste, and vast numbers of shorebirds were being decimated by market hunting. In *Wildlife of America*, a dramatic history chronicling the destruction of wildlife across the continent, Peter Matthiessen noted that during the last quarter of the nineteenth century, "the destruction of birds of all sizes and shapes had reached the proportions of a national pastime."

Although wildlife laws were developing in North America as early as the seventeenth century, they were frequently aimed at regulating game or "vermin" rather than protecting wildlife. It was not until the destruction of birds reached the proportions described above that truly significant and cohesive efforts to protect them were finally pursued. Those efforts began in earnest in 1883 with the establishment of the American Ornithologists' Union (AOU). An organization of mostly professional ornithologists, the union quickly formed a committee "for the protection of North American birds and their eggs, against wanton and indiscriminate destruction." In 1886, the committee proposed the adoption of a bird-protection law for each state, which came to be known as the "model law," the prototype for bird legislation in the United States. This law forbids the destruction of all nongame birds and specifies protection for nests and eggs of both game and nongame species.[3]

One particularly instrumental member of the AOU was the zoologist and conservationist George Bird Grinnell. The same year the model law was proposed, Grinnell organized the first US society for the protection of wild birds, naming it the Audubon Society.[4] At the time, Audubon's representation of birds as something more than simple commodities provided a new and important perspective, and the artist's name was chosen, Grinnell stated, because he had done more to teach Americans about birds and the need for their conservation than anyone else thus far.[5] Within two years, Audubon societies were established in sixteen states. At about the same time, the first bird sanctuary in North America was officially designated at Last Mountain Lake, Saskatchewan. By 1900, the model law, also known as the Audubon law, had passed in numerous states, and so had the Lacey Act, making it a federal crime to kill birds illegally in one state and sell them in another. In 1903, President Theodore Roosevelt acted on an initiative of numerous Audubon Society chapters and declared Florida's Pelican Island a federal bird reservation, a place that would be "a preserve and breeding ground for native birds." This action established the first National Wildlife Refuge (NWR) in the United States and created a federal system for protecting birds and other wildlife at places critical for their reproduction and survival.

As these developments were occurring, another important early AOU member, the businessman and self-made ornithologist William Dutcher, was collaborating with the naturalist and artist Abbott H. Thayer to establish a fund for the protection of seabird colonies along the Atlantic Coast. They hired wardens to protect birds at key sites, and by 1904 thirty-four wardens were employed under the Thayer Fund in ten states. In 1905, Dutcher was elected president of the newly incorporated National Association of Audubon Societies; sadly, the first murder of an Audubon warden occurred in the same year. As Guy Bradley tried to arrest plume hunters pursuing egrets at Oyster Key, Florida, he was shot and killed. Three years later, a second Audubon warden, Columbus G. Macleod, was murdered in Florida by plume hunters near Punta Gorda. These murders inspired public outrage and ultimately helped shift public and political support for the Audubon Society and efforts to protect birds.

By 1909, an additional fifty NWRs had been established across the United States. Canada had likewise established many migratory bird sanctuaries during the same period. In 1916, the United States and Canada signed the first international treaty specifically designed to protect migratory birds, and Congress enacted the Migratory Bird Treaty Act (MBTA) in July 1918. Recognizing that many birds migrating between the two countries were in danger of extermination, this treaty established protections from indiscriminant slaughter, making it a federal crime to kill migratory birds, destroy their nests or eggs, or possess any part of them. But the MBTA was established during a time when predators were regarded as "murderers," and the beneficent attitude toward migratory species did not encompass the entire group.[6] Therefore, those protected under the convention were only "such migratory birds as are either useful to man or are harmless," including game birds, insectivorous birds, shorebirds, seabirds, wading birds, pigeons, doves, and others. Conversely, the motley assortment still considered "vermin"—birds of prey, cormorants, pelicans, crows, ravens, and others that in one way or another impinged on human interests—were left off the protected list. Nevertheless, in this new environment of awareness and protection, even some of these transgressors would benefit, the cormorant being one of them.

Atlantic Coast

The New England coast was an early and particularly important area for seabird conservation. It was there that cormorants made their most dramatic initial recovery. The area, dominated by the coast of Maine, is characterized by numerous bays dotted with thousands of isolated islands, thus providing ample foraging and breeding habitat for seabirds. Although hunting for food and feathers had decimated many species in the region, the area's capacity to support large numbers of seabirds remained intact. Fortuitously, early twentieth-century efforts to protect species remaining on specific sites coincided with human emigration away from islands.

In 1931, a paper published in the *Auk* summarized changes in seabird populations along the Maine coast and directly attributed them to the efforts of the Audubon Society. The ornithologists Arthur H. Norton and Robert Allen wrote:

> For the last 27 years the National Association of Audubon Societies has employed wardens to guard the more important bird colonies on this coast. It is interesting to view the changes that have been wrought through the protection thus afforded. Herring Gulls have increased to such numbers that they are now a menace to the Terns, and in many instances have usurped their breeding grounds. The first definite records of the breeding of the Great Black-backed Gull, and the very great increase in the number of breeding Cormorants, are both significant facts.

The "great increase in the number of breeding cormorants" refers to the discovery in 1931 of five breeding colonies of double-crested cormorants, representing the first breeding records obtained in New England since the species had been extirpated in the 1800s. At the time of their discovery, Norton and Allen estimated that more than 1,700 breeding adults were present.

Exactly when cormorants recolonized New England is not known, but Norton made annual visits to the Maine coast between 1902 and 1923 and found no nests. Howard Mendall suggested that cormorants probably resumed breeding in the area about 1925. Based on the large number estimated in 1931, this is a reasonable assumption, since it likely took at least a few years to reach those numbers. Recovery along the coast then proceeded at a spectacular pace. By the late 1930s, breeding was once again documented as far south as Boston Harbor, and 10,000 pairs of double-crested cormorants were conservatively estimated to be breeding along Maine's coast in 1943. The first New England–wide population estimate, made in 1945, put the population at 13,000 pairs. By 1946, the population had extended farther south, and breeding was documented on the Weepecket Islands in Buzzards Bay, Massachusetts.[7]

Farther east on the coast in Canada, birds were likewise reclaiming former breeding areas. By the mid- to late 1950s, 1,300–1,500 pairs were estimated in Nova Scotia, where birds had been all but extirpated in the early 1900s. Though it is not known exactly when birds resumed nesting in this province, they are believed to have begun recovering around the same time as birds along the Maine coast. The origins of the birds recolonizing these areas remain a mystery. One hypothesis is that they immigrated to coastal Maine from eastern Canada, but the simultaneous recovery in Nova Scotia makes this assumption questionable.[8]

Quebec's north shore of the Gulf of St. Lawrence was another important area for early seabird conservation efforts. As a result of the Migratory Bird Treaty and the Convention Act in Canada, ten migratory bird sanctuaries were established on the north shore in 1917. Each consisted of two or more islands and had a part-time warden who was a local resident. For thirty years, the program was under the leadership of H. F. Lewis, who worked as the chief migratory bird officer for Ontario and Quebec from 1920 to 1943, became superintendent of wildlife protection for Canada in 1944, and was chief of the Canadian Wildlife Service from 1947 to 1952. Although the sanctuaries were initially established for other species, Lewis gave some idea of their early importance for cormorants. In his PhD thesis, he reported that outside the sanctuaries, cormorant colonies were annually raided to supply fresh meat for Indians, fresh eggs for local residents, and food for sled dogs and captive foxes. Those uncontrolled uses put a severe strain on many colonies, and some lost nearly all of their young every year.

Starting in 1925, efforts were undertaken to document seabird numbers and to assess the effectiveness of protection at the sanctuaries; censuses have since been conducted nearly every five years. Results are regularly published in the journal *Canadian Field-Naturalist*, and there is now an exceptional monitoring record, including the sixteenth census, conducted in 2005, which perpetuates an eighty-year-old tradition. The first four censuses, undertaken by Lewis, showed cormorants persisting at a few of the sanctuaries, in numbers ranging from approximately 800 to almost 1,400 birds. During the census of 1930, numbers of cormorants at the sanctuaries declined, a change Lewis attributed at least partly to killing that occurred outside the sanctuaries. Since that time, the number of cormorants in the sanctuaries has fluctuated, but they have consistently been found in at least a few of these places, indicating that the sanctuaries provided important refuges for cormorants throughout the twentieth century.

The Great Lakes: A Mysterious Appearance

Although published literature on early bird declines and conservation efforts in the waters of the Great Lakes is limited, there is enough documentation to determine that colonial waterbirds there experienced pressures similar to those affecting seabirds in the East. Commentary included by Jefferson Butler in *The History, Work and Aims of the Michigan Audubon Society* (1907) indicates the extent to which birds in general were persecuted in this area: "To read the full history of bird destruction in this state for one year would be akin to reading some of the Inquisition intrigues of the Middle Ages. One would not need to be a sentimentalist to have a feeling of nightmare." The gulls and terns of the region were relentlessly pursued for feathers, eggs, and food by plume hunters, Native Americans, and fishermen. The ring-billed gull, currently the region's most abundant colonial waterbird species, had essentially been extirpated as a breeding bird on Lakes Michigan and Huron. The ubiquitous herring gull had disappeared from many of its more southern sites and was restricted to the northern portions of the lakes. The Caspian tern, although never abundant, was headed for extirpation in Michigan waters. And the once-abundant common tern, by then "very greatly diminished," had become the opposite of what its name implied.

To combat these declines, specific actions were undertaken to protect important nesting islands. In 1905, the Midwest's first NWRs were established in Lake Superior on Michigan's Huron and Siskiwit Islands, both specifically for the protection of native birds. Concurrently, the Michigan Audubon Society began hiring wardens to protect gulls on islands in Lakes Superior and Michigan. By 1913, several islands in the Wisconsin waters of Lake Michigan had been recognized for their importance to colonial waterbirds, and the Gravel Island and Green Bay NWRs were established for their protection. Along with these measures, protection afforded to gulls and terns under the MBTA further discouraged visits to nesting islands by plume, market, and egg hunters. These combined actions resulted in significant protection for these species and the islands where they nested, and soon led to population increases.

Sometime between 1913 and 1920, ornithologists obtained the first definitive breeding records for double-crested cormorants in these waters; sites supporting ten to twenty pairs were observed in Lake Superior. But exactly when cormorants began nesting at these sites is not known. One of the first sites documented, Agawa Bay in eastern Lake Superior, Ontario, was visited by ornithologists in 1926, but local residents reported that cormorants had been nesting there "for years." By the late 1930s, observations of nesting cormorants were being documented by bird banders and ornithologists on all five Great Lakes. No formal surveys to assess cormorant numbers were undertaken until the 1970s, but some estimates of population size between the 1930s and 1950s have been proposed. In the late 1970s, the ornithologist Sergej Postupalsky reviewed breeding records from the late 1940s or early 1950s and estimated that close to a thousand pairs were then present in the Great Lakes, which he considered the period of peak population size. But in the mid-1990s, the ornithologist James P. Ludwig, whose family had been banding colonial waterbirds on the Great Lakes since the 1920s, reviewed cormorant nesting records with his colleague Cheryl Summer. Their review, which incorporated nest counts obtained during banding efforts for gulls and terns at islands on Lakes Michigan, Superior, and Huron in the 1930s and early 1940s, indicated that peak population size may have occurred earlier and been larger. Results suggested that 1,700–3,000 pairs may have been present in the upper Great Lakes in the 1940s, with most birds breeding in Canadian Lake Huron.[9]

Whatever the actual population size may have been during the 1930s and 1940s, population growth during this period was much slower than the concurrent growth in New England. Ludwig and Summer suggested that persecution by commercial fishermen and predation by gulls during nesting probably kept the population at a low level. A second important distinction between the Great Lakes and New England lies in how the cormorant's twentieth-century presence in each area has been interpreted. A well-documented record of the cormorant's demise in New England indicated that changes since 1900 clearly represented population recovery. But no such record for the Great Lakes exists, making the interpretation of the cormorant's twentieth-century presence in those waters less clear. Considering the current significance of this region to cormorants, and the extensive persecution this species has experienced there, the question of how the cormorant's presence in the Great Lakes has been interpreted deserves a more detailed discussion.

Before the twentieth century, breeding cormorants were documented in most of the states and provinces bordering the Great Lakes, and they were certainly part of the Great Lakes watershed.[10] But because there are no records of nesting on the Great Lakes themselves, the observations documented between 1913 and 1920 on Lake Superior are generally regarded as the time at which cormorants colonized the Great Lakes. The first colonies are thought to have developed in western Lake Superior, and a pattern of colonization proceeding from the northwest to the southeast is presumed to have occurred. The birds were described as "newcomers," and their occurrence in those waters was considered a range expansion or, more to the point, an "invasion."[11]

But whether cormorants were really "invaders" has been a matter of some debate. Why cormorants would not have been breeding there before the twentieth century poses an intriguing question. With many thousands of islands and abundant fish resources, the Great Lakes contain ideal habitat for these birds. Given the resources available, and the distribution of cormorants in states and provinces bordering the Great Lakes, the cormorant's absence from the area is mysterious indeed. Why would cormorants have avoided some of the highest-quality habitat in the middle of their breeding range?

One possible explanation is that the fish community may not have been as attractive to cormorants as it later became. By 1900, overfishing and other human activities had caused significant changes in fish species composition, and better foraging opportunities for cormorants may have resulted. But fish-eating gulls and terns had thrived on the lakes long before the fish community was heavily affected by human activities. The presence and abundance of those birds makes it highly doubtful that the cormorant—the fishing bird in this community with the greatest ability to pursue diverse fish species through the widest range of water depths—could not have been supported in at least some numbers in this environment.

Another possibility is that cormorants, once present, had disappeared before breeding could be documented. Limited archeological evidence from other locations in the cormorant's range indicates that cormorants were present hundreds or even thousands of years ago, but disappeared from these places long before people began collecting records to document breeding distributions. There are good examples of this on the Atlantic and Pacific Coasts. For the Great Lakes, no archeological evidence indicates that cormorants were present as breeding birds. But several other kinds of records (anecdotes, geographic place names, accounts of general breeding distribution, checklists) identified in reviews of the species' historical distribution indicate that cormorants occurred in or near these waters before the twentieth century. At certain locations, they had undergone significant declines before they were documented through formal observations of breeding.[12] Some of these records challenge the timing and pattern of presumed colonization in this area. Furthermore, since these reviews were completed, additional records of a similar nature have been identified, and more could likely be found through a focused search of unpublished historical documents.[13]

In accounting for the cormorant's absence from the Great Lakes, another important consideration is the documented persecution and demise of other colonial waterbirds in the region. As a group, colonial waterbirds are particularly sensitive to intrusions at breeding sites, and repeated colony raids would have been detrimental to all species. But cormorants may have been at a particular disadvantage, since they almost always share sites with gulls, which opportunistically raid the nests of other colonial waterbirds when human disturbances occur. The resulting loss of eggs and chicks to gulls, in addition to humans, would have had a massive reproductive impact on cormorants. Such impacts from gull predation arising from human disturbance were observed in other areas, and under such pressure, cormorants could have declined earlier than gulls and terns, just as they had on islands on the East Coast. Likewise, as gulls and terns became protected and began to recover, cormorants would have also benefited.

While the cormorant's past in the Great Lakes remains mysterious, the birds were, without a doubt, widely present in the surrounding basin before they were first documented breeding in the lakes themselves. Whatever their former breeding status, cormorants were positioned to respond almost immediately to the protection afforded to other birds in the area. Their early twentieth-century presence on Great Lakes islands coincides with the region's protection of other colonial waterbirds, a pattern observed elsewhere on the continent. From this perspective, their seemingly sudden "invasion" should be of little surprise.

The Lake Winnipeg Watershed

During the late nineteenth and early twentieth centuries, the Lake Winnipeg watershed was a unique area within the cormorant's continental distribution. The watershed, the second largest in Canada, drains much of the prairie region in the central portion of the continent, including parts of Alberta, Saskatchewan, Manitoba, and Ontario, as well as portions of Minnesota and North Dakota. Distributed throughout are many large, island-studded lakes that form

an important core breeding area for cormorants. Unlike many other portions
of the range, several of these lakes continued to support large colonies during
the period of intense persecution. H. F. Lewis estimated that there were some
8,000 breeding birds in the early twentieth century on Lake Manitoba, Lake
Winnipegosis, Last Mountain Lake, and Little Pelican Lake, and that numbers in the region made up nearly one-third of the entire interior and Atlantic
population. Moreover, the actual proportion was probably greater because of
the difficulty in locating all cormorant colonies in areas such as Manitoba and
Saskatchewan.[14]

Many of these remote interior locations are now renowned for their commercial fisheries of walleye and other freshwater species. The importance of
these fisheries to humans is relatively recent; significant commercial fisheries
on Lake Winnipegosis and Lake of the Woods did not develop until the late
1800s or 1900. But their rich fish resources and abundant island habitat have
long been attractive to fish-eating birds. Breeding records for cormorants go
back to the late 1700s, when chicks were observed at Lake of the Woods, but
they were undoubtedly present there long before records were kept.

Given the area's remote location, abundant resources, and relatively low
human presence, it is not surprising that breeding birds persisted there and
recovered in significant numbers in at least some locations. By the mid-1940s,
the population at Lake Winnipegosis was estimated at nearly 10,000 pairs. No
lake-wide surveys had been conducted there previously, but information compiled from observations of biologists and local fishermen suggests that cormorant numbers increased rapidly up to 1943. At Lake of the Woods, the increase was better documented, going from just 50 pairs in 1927 to an estimated
4,000–6,000 birds by 1946. No sanctuaries were established at either location
or, for that matter, anywhere in Manitoba, suggesting that birds recovered at
those locations without the benefit of direct protection. Possibly, birds surviving elsewhere because of protection from human persecution immigrated to
these lakes and contributed to rapid increases.[15]

At some other locations, conservation efforts were likely important. In
Saskatchewan, the Last Mountain Lake sanctuary was one of the few locations
in the region to support large cormorant colonies in the late 1920s. Johnson
Lake, Old Wives Lake, Big Quill Lake, Little Quill Lake, and Redberry Lake,
all in Saskatchewan, were sanctuaries by the late 1920s, and cormorants persisted at many of these places. In Alberta, sanctuaries were established at Lac

La Biche and Miquelon Lake, both locations where cormorants persisted in fairly large numbers. Conversely, Stump Lake in North Dakota was designated an NWR in 1905, but cormorants were still being persecuted there in 1925.

Florida and Its Coasts

Much of the Florida portion of the cormorant's range is dominated by a nearly impenetrable maze of mangrove and cypress swamps, and historical information is sparse. Limited observations suggest that while cormorants remained abundant in Florida, they experienced local declines. Audubon stated that the cormorant was "constantly resident in the Floridas and their Keys, and along the coast to Texas," but by the twentieth century, cormorants were known to breed along the Gulf Coast only in Florida and Louisiana. On the Atlantic side, coastal North Carolina was the only other southern location where cormorants had been recorded as an abundant breeder. According to the historian John Brickell, cormorants had formerly nested on the islands in "the Sound" and near the shore in "the Banks" (see chapter 3), but by the late 1800s, only one small colony was documented, on coastal Great Lake in Craven County. This was the only known breeding location for North Carolina until the 1950s.

Although not a primary target of plume hunters, cormorants were likely affected by raids on mixed colonies to obtain feathers from wading birds. Additionally, Audubon and others recorded the persecution of pelicans by fishermen in this region, and co-occurring cormorants would have certainly been targeted at the same time. Audubon reported seeing cormorants for sale as food in the markets of New Orleans, although the origin of those birds was not clear. Because of plume hunting, wading-bird colonies in Florida and Louisiana became protected by wardens, and many of the first national wildlife refuges were established in those states for wading-bird protection. Cormorants likely benefited at some of those sites simply because intrusions into colonies were limited. Additionally, nesting on small, mostly inaccessible mangrove islands probably contributed to their persistence in parts of the region.

Alaska and the Pacific Coast

On the western portion of the continent, a range of complex factors limited the extent to which conservation actions allowed cormorants to recover. In Alaska, the Aleutian Islands were designated as an NWR in 1913 and set aside for the protection of native birds. But they were reserved also for the propagation of fur-bearing animals, and fur farming peaked in the 1920s. How much the introduction of such mammals affected cormorant populations is not clear, but in the mid-1990s, large and diverse seabird populations were reported to exist only on a few islands between southeastern Alaska and the Aleutian Islands where foxes had never been introduced, indicating that impacts to all seabirds were substantial.[16]

In the Pacific Northwest, the first nesting record from the inner coastal waters of British Columbia was obtained in 1927 on Mandarte Island in the Strait of Georgia. Protected from human disturbance since 1914, the site probably provided a safe haven for cormorants. In 1933, a second colony in the same area, thought to be the older of the two, was discovered at Ballingall Islets. In 1937, two more colonies were discovered in the vicinity of the nearby San Juan Islands in Washington. These four colonies were the only known sites in the inner coastal waters until the late 1940s. While insignificant numerically, they deserve special note because they indicate important events in the cormorant's history. In 1989, the researchers Keith Hobson and J. C. Driver published an analysis of bird remains excavated from archeological sites around the Strait of Georgia, revealing that double-crested cormorants had been present between 3500 BC and AD 1800. But because no bones of young birds were retrieved, the researchers could not determine whether birds had been breeding in the area, so it is not clear whether twentieth-century records represent recolonization. For the San Juan de Fuca area, however, the archeological analysis by Kristine Bovy referred to earlier definitively showed that the early twentieth-century records indicated recolonization.

Along Washington's outer coast, three NWRs, Copalis, Quillayute Needles, and Flattery Rocks, were established in 1907. They encompass significant stretches of the coast and many islands and rocks important for hundreds of thousands of nesting seabirds, and cormorants have persisted at several of these locations. The West Coast's first NWR was established in 1907 off the coast of Oregon at Three Arch Rocks, a collection of nine rocks that provided habitat for abundant concentrations of several seabird species. This location was set aside specifically to protect seabirds from what had become a popular Sunday pastime: boatloads of shooters went to the rocks to enjoy a day of target practice and sport, and shooting parties killed thousands of seabirds at a time.[17] Protection was likely integral to the site becoming the location for the largest cormorant colony on the Oregon coast in the 1970s. But at other nearby coastal locations, raids on cormorant colonies continued and may have prevented numbers from increasing. The Oregon Islands NWR, established in 1935, afforded protection for nesting seabirds and other marine animals on hundreds of islands, shoals, and rocks along Oregon's coastline.

California's Farallon Islands were designated an NWR in 1909, but active wildlife protection did not begin until the 1970s. For numerous reasons, double-crested cormorants did not recover their former abundance there. In the 1940s, the Pacific sardine fishery declined and ultimately crashed, a major ecological event in the offshore waters of California and Baja California that has been linked to the double-crested cormorant's failure to recover along this portion of the coast.[18] Finally, in the interior, NWRs were established between 1908 and 1911 at Clear Lake, California, and Klamath and Malheur Lakes in Oregon. Early reports indicated that cormorants persisted at those locations specifically because of protection efforts.

5

Reversal of Fortunes

Another Decline and the Second Recovery

It developed that everyone in Cape San Lucas hates cormorants.
They are the flies in a perfect ecological ointment. . . . They dive
and catch fish, but also they drive the schools away from the pier
out of easy reach of the baitmen. Thus they are considered inter-
lopers, radicals, subversive forces against the perfect and God-set
balance on Cape San Lucas. And they are rightly slaughtered, as
all radicals should be. As one of our number remarked, "Why,
pretty soon they'll want to vote."

—John Steinbeck, *Log from Sea of Cortez* (1951)

WHILE GREAT STRIDES toward bird conservation were made during the
first half of the twentieth century, attitudes toward cormorants continued to
reflect those of times gone by. Steinbeck's observations describe a midcentury
mindset that prevailed not only at Cape San Lucas but almost anywhere cormorants could be found. From Baja to Quebec, the story was the same: as
cormorant populations recovered or expanded, the old hatred reawakened or
a new one was born. In his tongue-in-cheek comment, Steinbeck identifies an

Cormorants on ground nests

important and lasting role reversal that distinguishes human-cormorant relations. Cormorants, returning to areas where they had fished in the past, or attempting to use unexplored or modified portions of their range, came to be seen as vagabonds intruding where they did not belong. Conversely, the colonizers from Europe had settled across much of North America, and their generations now characterized the lands. In many places, the cormorant returned to its historic or potential breeding grounds only to find the waters appropriated by commercial and sport fishermen, who harvested millions of fish annually and would not tolerate competition from these interloping radicals.

Almost immediately after cormorant colonies began to reestablish, attacks on cormorants resumed. In four areas in particular, the Lake Winnipeg watershed, the New England coast, the Great Lakes, and the St. Lawrence estuary, persecution became so widespread that it developed into a regional rather than a local phenomenon. This period marks yet another significant turning point in the species' history. As the cormorant recovered, the persecution that occurred in each of these regions led to the establishment of the first government-sanctioned control programs. Finally, an official policy for dealing with this subversive force of nature was born.

The Fly in the Ointment: The Official Approach

The first such program was initiated in the Lake Winnipeg watershed. At Lake Winnipegosis, where cormorants had always experienced some degree of abundance, commercial fishermen had sporadically destroyed large numbers of eggs and young on numerous islands for years. In the 1940s, fishermen reported rapid increases in cormorant numbers and lodged numerous complaints to the Manitoba Game and Fisheries Branch. In response, the agency initiated a diet study and launched an official cormorant control program in 1943. Although cormorants were eating mostly rough fish, with commercially valuable fish species constituting only about 7 percent of their diet, their consumption of commercially valuable fish was considered too large. During the first three years of the program, the Manitoba wildlife personnel visited all known nesting islands during the height of the breeding season and destroyed all eggs and young birds encountered. In addition, fishermen and loggers continued to destroy cormorants.[1]

Over a six-year period, these efforts resulted in the destruction of many thousands of eggs and chicks, reducing the Lake Winnipegosis population by more than half. The first complete count of all known nesting islands, conducted in 1945, estimated that there were 9,862 nests and 39,448 adults. By 1951, the number of nests had declined to 4,656, and the number of adults to 18,624. The Manitoba Game and Fisheries Branch considered this population level to pose a less acute threat to the fishery, and therefore the official control program ended.[2]

One year after control was initiated on Lake Winnipegosis, the first formal control program for cormorants in the United States was launched by the US Fish and Wildlife Service (USFWS), along the coast of Maine. In this region, fishermen using fishing weirs began reporting conflicts with cormorants in 1934, just three years after the cormorant's return as a breeding species was documented. By the early 1940s, adult cormorants were again being shot, and eggs and nests destroyed. By 1944, Maine's fishing industry had lodged multiple complaints of alleged cormorant depredations on fish, prompting the USFWS to take action.

Over the next ten years, a program of egg oiling was pursued on numerous islands along Maine's coast, a practice that allows eggs to remain intact but cuts off the oxygen supply to developing chicks. Because adults typically re-lay when eggs are lost, this method tricks adult birds into remaining and incubating eggs that will never hatch, thereby forfeiting the chance of producing young that year. More than 187,000 cormorant eggs were sprayed with oil during this period, but the breeding population did not decline. In fact, the year before the program was initiated, the breeding population was conservatively estimated at 20,000 birds, while ten years later, at the program's end, the population was estimated at no less than 22,470 individuals.

In 1953, the Maine Department of Sea and Shore Fisheries assumed responsibility for the program and undertook initial research to determine the effectiveness of egg oiling. In its assessment, the agency noted that there was no evidence that the program was controlling cormorant numbers; at best, it appeared to be simply slowing population growth. Notably, reducing reproductive success is a slow approach to controlling population size in a long-lived bird, possibly requiring several years to show results. But the agency's assessment identified several more immediate reasons for the program's failure, including inadequate coverage throughout the breeding season, breakage of oiled eggs, and a limited understanding of cormorant behavior. Considered largely ineffective, the program was discontinued the following year. Moreover, the agency stated that there was no definite evidence to indicate that the population should even be controlled, and recommended that the economic influence of cormorants on commercial fisheries be investigated.[3]

The next formal control program initiated during this period was launched on the Great Lakes in Ontario waters. In this region, like everywhere else, conflicts with fishermen occurred almost as soon as cormorant colonies were discovered. By the early 1940s, reports of cormorants being shot and their nests and eggs destroyed on several islands in both US and Canadian waters were numerous. But the Great Lakes region was different from other areas in that much smaller cormorant numbers incited fishermen into action, probably because the birds were considered "newcomers" to the area. Tolerance for them was next to nil, as evidenced by the initiation of a government control program when at most 3,000 pairs (and possibly far fewer) were present in all the Great Lakes.

Specifically, fishermen in the North Channel and Georgian Bay areas of Lake Huron, Ontario, provided the impetus for control. In the late 1940s and early 1950s, each of these areas supported a few hundred pairs of cormorants on a handful of rocky islands. Persecution by fishermen was especially intense, and at nearly all known colony sites, birds and nests were destroyed. In the mid-1940s, commercial fishermen from these areas made complaints to the Ontario Department of Lands and Forests (now the Ontario Ministry of Natural Resources), alleging that cormorants were detrimental to black bass and sport fish as well as to pound-net fishing. Impacts on sport fish and black bass were examined in 1946 by the ornithologist James L. Baillie, Jr., to investigate "economic relations" associated with cormorants in Georgian Bay. Baillie's study revealed the consumption of only one small black bass, indicated that the local ranges of cormorants and black bass did not overlap to any significant extent and that cormorants were consuming species unimportant as sport fish. Baillie concluded that "the black bass fishing, for which the . . . region is famous, was poor in 1945 and 1946, and it seems that the cormorants were made the scapegoats." Despite Baillie's report, in 1946 the Ontario Department of Lands and Forests authorized the first Great Lakes control program, focused primarily on Georgian Bay. Egg destruction was used first, followed by egg oiling. Meanwhile, illegal activities by fishermen continued. The official cormorant control program remained in effect until 1966, but was not considered effective or intensive enough to have reduced population size. As on the Maine coast, control efforts were believed only to have slowed population growth in the region.[4]

The last formal control program initiated during this time period was launched in Quebec by the Canadian Wildlife Service (CWS) on Île aux Pommes in the St. Lawrence estuary. In this region, fishermen were concerned about impacts on fish, and local residents were concerned about impacts on trees from cormorant nesting activities. In 1954 and 1955, CWS officers killed all chicks on Île aux Pommes and treated eggs with substances to suffocate developing embryos. Return visits to the island indicated that adult cormorants were still incubating the dead eggs. But after two consecutive years of control, the number of nests the following year did not decline, and so the program was discontinued.[5]

Elixirs of Death: Some New Influences on Cormorants

At about the same time these official cormorant control programs were getting under way, a new and more insidious influence was about to shape the cormorant's story for the next several decades. In 1939, the Swiss scientist Paul Hermann Müller discovered the insecticidal properties of DDT, an organochlorine compound that had great value in controlling insect pests that spread diseases such as malaria and yellow fever. By the middle of 1944, the US armed forces began using DDT as the standard insecticide in the battle against disease; in 1945, DDT and several other pesticides were made available to farmers as agricultural insecticides. The compound was immediately put into wide use as a "miracle powder," and the DDT era had begun.

All organochlorines are slow to degrade, persisting in the environment for a long time after they are initially deposited. Because DDT is insoluble in water but highly soluble in fats, it has a great affinity for the tissues of living organisms. In most animals, the chemical breaks down to DDE, which tends to stay in the body and increases in concentration tremendously as it moves through the food chain. The later discovery that DDT was directly toxic to fish and crabs, and indirectly to a host of other animals, led Rachel Carson, in her groundbreaking book *Silent Spring*, to call it one of several "elixirs of death." In birds, high concentrations of DDE result in a variety of problems. In some species, one result included ineffective calcium metabolism, leading to an inability to form sufficiently thick eggshells. This effect was best exemplified by the drastic declines that occurred in bald eagle, brown pelican, osprey, and peregrine falcon populations. Soon after DDT was introduced, each of these birds began producing eggs with shells that were too thin to support their weight during incubation, with disastrous population-level impacts.

Although not as well publicized, dramatic effects on cormorants across much of their range also occurred from DDT and other contaminants. On the Pacific Coast, eggshell thinning and declines in reproductive success observed in the late 1960s at colonies in Southern California and northwest Baja California were attributed to the presence of a DDT manufacturing company in Los Angeles.[6] Along the Maine coast and elsewhere in New England, the rapid growth that cormorants had experienced up to 1945 came to a halt, which was not attributed to the official control program. Growth remained very slow for the next two decades, and studies published in the 1960s and early 1970s found DDE present in cormorant eggshells at several locations along the coast. Illegal control of cormorants also continued through this period, but reports of DDE contamination, poor fledging success, and the timing of slow growth suggest that DDT may have been a more important factor.[7]

Similarly, eggshell thinning was documented across Canada, and varying levels of DDE residues were present in cormorant eggs in all provinces. At remote Lake Winnipegosis, the cormorant population continued to decline even after the official control program ended. By 1969, just 1,403 pairs were present, a decline of 70 percent since 1951, the last year of official control. While fishermen continued to destroy eggs, nestlings, and adults during this period, the fact that the population there increased during other periods of illegal control suggests additional factors may have contributed to the decline. Lake Winnipegosis receives substantial runoff from agricultural areas, and DDE residues were documented in eggs there. Thus, pesticides may have suppressed repro-ductive success and contributed to diminishing cormorant numbers even at that seemingly pristine site.[8]

By far the most significant impacts were documented on the Great Lakes. As early as 1955, cormorant eggs collected on Lake Michigan had shells that had thinned by as much as 10 percent, and by 1959–1961, eggs collected from Ontario waters had shells that had thinned by as much as 27 percent. Simulta-neously, cormorant numbers were decreasing in several areas, including Lakes Michigan and Superior and the Georgian Bay area of Lake Huron. Sometime between 1960 and 1962, the cormorant disappeared as a breeding species from Michigan, while in neighboring Wisconsin a remnant thirty pairs hung on at just four sites on inland flowages. By 1970, only eighty-nine pairs were esti-mated across the entire Great Lakes, and on nearby Lake Nipigon, a total of just twenty-seven pairs persisted. In 1972, 95 percent of eggs observed in col-onies on Lake Huron had broken or disappeared by the end of the incubation period, and Great Lakes cormorants were considered the most heavily con-taminated cormorants in North America, their eggshells averaging 24 percent thinner than pre-DDT values. While both legal and illegal cormorant control continued during this period, multiple lines of evidence point to DDT as a more significant factor in the declines.[9]

But DDT was not the only chemical in the mix. At least for cormorants nesting in the center of the continent, it is likely that other contaminants played a role in their precipitous decline in the 1950s. Early in the twentieth century, very large numbers of cormorants migrated along the Mississippi River in spring and autumn. These birds probably bred in the upper Great Lakes and on Canadian lakes to the north and perhaps as far west as Lakes Winnipeg and Winnipegosis.[10] Migratory flights along the Mississippi were said to have numbered in the millions in 1820, hundreds of thousands in the 1920s, and tens of thousands in the 1940s and 1950s. Numbers decreased in the late 1950s and were virtually gone by 1964.[11] The colonial waterbird expert Ian Nisbet pointed out to me that this decline coincided exactly with the population crash of brown pelicans in Louisiana, whose numbers decreased from 20,000 in 1958 to 0 in 1963. It is generally agreed that the abrupt crash in the pelican popula-tion was caused by poisoning from endrin, a chlorinated pesticide that is much more toxic to birds and mammals than DDT. During the 1950s, endrin was discharged into the Mississippi River from a manufacturing plant in Memphis, Tennessee, leading to massive fish kills downstream and lethal concentrations in fish in the Mississippi Delta and along the Gulf Coast to the west. For a fish-eating bird migrating down the Mississippi River that stopped to feed, the lower reaches of the Mississippi must have been a death trap. The decline in the numbers of cormorants seen on migration was much too rapid to have been caused by impaired reproduction alone. Some factor leading to increased mortality of adults appears to have been responsible in addition.[12]

Tourism, Another Factor in the Decline

At the tail end of the DDT era, ornithologists documented the demise of the greatest cormorant colony ever reported, an event attributed to yet a third influence that developed rapidly during this period. In 1969, when Isla San Martin was visited during the contaminants study, the huge cormorant colony that had existed in the early part of the century was estimated at just 5,000 pairs. Gone were the hundreds of thousands of cormorants described by Wright, and other seabirds on the island were also greatly diminished. Exactly when these declines occurred is not known, but the most important associated factor appears to have been the keen interest in seabirds that began developing concurrently with cormorant control and the heavy use of pesticides. By 1972, tourism had increased significantly on the west coast of Mexico, and educational tours during the breeding season were occurring almost weekly on Isla San Martin. Although this activity had the potential to increase awareness and understanding of fish-eating seabirds, many guides either did not understand or care about the sensitivity of birds to disturbance during the breeding season. The intense human disturbance had disastrous effects on nesting seabirds. By 1977, a visit by ornithologists revealed that no cormorants were nesting on San Martin, and people were observed all over the island.[13]

Protection from People and Contaminants

During this second phase of cormorant decline, public attitudes toward wildlife, the environment, and conservation continued to evolve. In 1949, two years before Steinbeck published *Log from Sea of Cortez*, Aldo Leopold published *A Sand County Almanac*. This collection of beautifully written essays articulated a conservation ethic that eventually promoted a giant step forward in human relations with wilderness and the land. Several essays proved to be milestones, and one in particular, "Thinking like a Mountain," introduced revolutionary ideas about the ecological role and value of predators. Reflecting on the death of an old wolf by his own hand, Leopold condensed a lifetime of study in three short pages. Like the power behind Audubon's paintings, Leopold's words captured the wolf as a fierce, defiant, and critical force within the landscape, and one whose death by human hands represented a loss not just of an individual but one for an entire mountain. Leopold reinvoked Thoreau's dictum "In wildness is the salvation of the world" and offered an entirely different context in which predators could be understood.

A little more than a decade later, Rachel Carson published *Silent Spring*, not only documenting the long-term environmental effects and threats due to pesticides, but also challenging the government and calling for a change in the way humans viewed the natural world. In 1964, just two years after this work was published and the year of Carson's death, the Wilderness Act was signed into law. By identifying wilderness itself as something worthy of preserving, the act has emerged as a landmark in the history of America's attitudes toward wild places and creatures.

By the late 1960s, the environmental movement was gaining momentum around the world. In Canada, most uses of DDT had been banned by 1969, and in 1971 a small team of environmental activists from British Columbia formed the nongovernmental organization Greenpeace, which went on to become the largest direct-action environmental organization in the world. In the United States, the National Environmental Policy Act (NEPA) became law in 1969, promoting a policy of environmental enhancement and requiring all federal agencies considering actions with the potential to affect environmental quality to first prepare an environmental assessment or impact statement. The following year, President Nixon established the US Environmental Protection Agency (EPA) to guide and oversee the enforcement of environmental policy. By 1972, DDT had been banned, and amendments to the Clean Air Act and the Federal Water Pollution Control Act imposed strict standards and controls to protect environmental quality. Additionally, a significant revision to the Federal Insecticide, Fungicide, and Rodenticide Act transferred responsibility for pesticide regulation from the US Department of Agriculture to the EPA. Finally, perhaps the single-most powerful environmental symbol ever to emerge was captured in that year when a photograph was taken by the crew of the last manned lunar mission, Apollo 17. From some 28,000 miles in space, the crew photographed the earth and provided a perspective of the planet never before seen. Amid an immense black cosmos, the entire globe is revealed, showing a lonely blue planet, small and vulnerable like an island in the sea. The power of this iconic image, known as the *Blue Marble*, spoke to the growing awareness of the earth's fragile nature and came to symbolize the environmental movement.

Against this background, the precipitous declines in cormorant numbers were recognized as part of a conservation crisis affecting a wide range of birds. In the United States, this recognition led to the first significant conservation actions ever undertaken specifically for cormorants. The year 1972 was indeed a magic one, when actions occurred that would alter the course of the cormorant's future for years to come. The cormorant was added to the National Audubon Society's "Blue List," a watch list that identifies North American species experiencing significant population declines and range reductions. In addition, the State of Wisconsin added the cormorant to its list of endangered species in an effort to protect the state's last four colonies. Of the greatest importance, however, was an amendment made to the Migratory Bird Treaty Act (MBTA) between the United States and Mexico on March 10, 1972. Prompted by the midcentury declines seen in so many species, the two countries signed an agreement to add an additional thirty-two families of birds to the protected list. Among those included were the predatory birds previously considered harmful: eagles, hawks, owls, corvids, pelicans, seabirds, and all members of Phalacrocoracidae.

Until then, the only real protection those birds received in the United States occurred when they were found on particular protected sites, such as refuges or sanctuaries, or in states with wildlife laws that covered them. In the latter case, protection was far from consistent and generally ineffective. For instance, cormorants were legally protected under state law in Wisconsin, but extreme persecution and vandalism were well documented at several colonies between 1956 and 1963.[14] In Maine and Minnesota, cormorants lacked any legal protection at all and could be (and were) killed without restriction. But with the addition of cormorants to the MBTA protected-bird list, the raiding of colonies to destroy adult birds, chicks, eggs, and nests became a federal crime, and cormorant protection became the federal responsibility of the USFWS.

In the next several years, the cormorant's diminished status received special attention from additional Great Lakes states. Besides listing it as endangered, Wisconsin began a cormorant restoration program in the mid-1970s, building artificial nesting platforms attached to telephone poles at the southern end of Green Bay. By 1984, 1,269 such platforms had been erected at thirteen locations.[15] In 1978, Illinois designated the cormorant as an endangered species and also provided artificial nesting platforms. In 1980, Michigan likewise assigned the cormorant endangered status.

In Canada, efforts to protect cormorants were also undertaken during this period, but unlike the United States and Mexico, Canada did not add cormorants to the MBTA protected bird list. Instead, cormorants remained the responsibility of provincial jurisdictions, several of which had established legal protection for the birds before 1972 under game and nongame wildlife acts. Like the states, the provinces varied in their approach to cormorants and in the timing of the protections they offered. For instance, Ontario had granted cormorants some legal protection as early as 1927, when birds, but not nests or eggs, were included in a clause added to the Game and Fisheries Act to protect most nongame birds. In Manitoba, cormorants did not receive any provincial protection until 1963, when they gained protection under Division 6 of the Wildlife Act. But as in the United States, such protection was largely ineffective, and persecution by fishermen continued long after the cormorant was given protected status. In 1966, the ornithologist Ralph Carson published a paper to "call attention to the wanton destruction of pelican and cormorant nests" in a colony on Suggi Lake, Saskatchewan, and to advocate for effective protection of these already legally protected birds. Carson recounts his 1964 visit to this colony, during which he observed fishermen camped out on the island and all cormorant and pelican nests destroyed. Upon his return to headquarters in Regina, Carson reported the destruction to an official with the Department of Natural Resources in charge of the Suggi Lake area. Despite the birds' legally protected status, the official condemned the pelicans "as a serious menace to the fishing industry and further stated that no sanction would be given to the birds."[16]

But as cormorant populations continued to decline and awareness of the cormorant's diminished status increased, efforts to obtain greater protection in many parts of Canada were made. Between 1968 and 1971, the ornithologist Kees Vermeer published a series of papers documenting declines and persecution of cormorants and pelicans across the Prairie Provinces. Vermeer continually argued for their conservation, and by 1971 several colonies had been given greater protection. In Alberta, declines in cormorant numbers led to the species being provincially designated endangered in 1977. Similarly, several of the migratory bird sanctuaries along Quebec's North Shore of the Gulf of

St. Lawrence became the focus of more vigorous conservation efforts in the early 1980s, and teams of wardens were established to regularly patrol the sites and protect nesting birds.[17]

Other factors may also have been instrumental in diminishing the persecution of cormorants. The environmental movement in general brought a new appreciation for ecology and its components to at least some members of the public. Further, cormorant numbers were so diminished in some areas that they appeared to be of little concern to fishermen. For instance, so few birds remained in the Great Lakes by 1970 that fishermen in most areas would have been hard-pressed to find any to kill. Additionally, during the period when cormorants were declining, overfishing and other factors greatly diminished yields of commercial fish species. By the 1970s, the number of commercial fishermen in the Great Lakes had diminished, and many of the commercial fishing licenses had been bought by natural resource agencies in an effort to help reduce the number of fish being removed by commercial fishermen. Moreover, concurrent changes in the fish community resulted in changes to cormorant foraging strategy. James Ludwig documented that cormorants were concentrating their fishing in depths other than those where fishermen's nets were being set. In addition, Ludwig observed that changes in commercial fishing gear further minimized direct interactions between commercial fishermen and cormorants. The decline in the use of gill nets was particularly important in diminishing interactions because cormorants had frequently become entangled in and damaged the nets.[18]

In sum, a combination of factors in the late 1960s provided conditions in which cormorants could again begin to recover. Declines in contaminants and the restoration of aquatic environments resulted in high-quality habitats once more being available, removing a major impediment to reproductive success. Although varying degrees of persecution continued in several areas, the new wave of protection and environmental awareness, along with less incentive to persecute birds in general, provided many locations where cormorants could raise chicks undisturbed by humans. This reduction in human disturbance distinguishes the recovery during the last third of the century from that which occurred during the first half; population growth in many areas proceeded largely unchecked by humans for the first time in hundreds of years. Furthermore, the declines that had occurred in bird populations during the pesticide era, coupled with a greater awareness of ecological vulnerability, resulted in

the establishment of systematic monitoring programs for waterbirds, which are considered sentinels of environmental change. By the mid- to late 1970s, efforts were in place in many areas to carefully track populations and document the truly spectacular recovery in cormorants that was about to begin.

There and Back Again

Over the next three decades, the cormorant's potential for rapid population growth began to be realized. In breeding areas that had had significant populations in times gone by, particularly dramatic increases began to occur. Along the northeastern Atlantic Coast, several areas underwent remarkable changes. In Quebec, substantial increases occurred in the St. Lawrence estuary, the North Shore, the Gaspé Peninsula, and the Magdalen Islands, with combined numbers in those areas growing from about 12,000 pairs in the late 1970s to 27,700 pairs in the late 1980s. In Canada's Maritime Provinces, similar increases occurred. Numbers in Nova Scotia went from 4,150 pairs in 1971 to 15,700 pairs in 1985. For Prince Edward Island and New Brunswick, estimates for the 1970s were not available, but by the mid- to late 1990s the Prince Edward Island population had reached close to 10,000 pairs, and 15,000 individuals were estimated along New Brunswick's coast during summer months. The population in New England saw dramatic changes, growing from 17,000 pairs in 1977 to nearly 38,000 pairs by the early 1990s. Additionally, the first formal nesting records for Connecticut, Rhode Island, New Jersey, and coastal New York were obtained during this period, probably because the *auritus* subspecies was spreading south.[19]

In the interior, even more spectacular changes were documented. Immense numbers amassed in the Prairie Provinces, with the population in Manitoba going from 4,772 pairs in 1969 to 22,681 pairs by 1979. At Lake Winnipegosis, the population rebounded from just 1,400 pairs to more than 9,000 pairs during the same period. Rapid growth continued, and within ten years the population had increased by more than 500 percent, reaching its known all-time high in 1989 with 51,788 pairs nesting on forty islands. Ten years later, the population had declined, returning close to 1987 levels, with about 36,200 pairs nesting on thirty-three islands. Saskatchewan's cormorant population underwent significant changes, growing from 1,078 pairs in 1968 to almost 11,000 pairs by 1982. Growth continued, and by 1991, the population had reached nearly 20,000 pairs, and by 2006 it was estimated at just over 34,000 pairs. In Alberta, cormorants resumed breeding at Lac La Biche in the 1970s, and numbers in the region reached over 8,000 pairs by 2005.[20]

On the Great Lakes, another exceptional story unfolded. By the mid-1970s, breeding birds were increasing in Canadian waters, and had returned in small numbers to Lake Michigan. The latter represented both the first known occurrence of breeding cormorants in Michigan since the early 1960s and a re-colonization of Wisconsin's Great Lakes waters. As early as 1980, cormorants were once again nesting on Lakes Huron, Michigan, Ontario, and Superior, and the population, which was represented by a mere 89 nests in 1970, was estimated at more than 1,300 pairs.

Curiously, during this initial period of recovery from the effects of DDT and other organochlorine contaminants, a range of physical deformities affecting multiple waterbird species began to be observed. In cormorants, the most famous of these was cross-bill syndrome, in which chicks hatched with bills that could not properly close, leaving them incapable of effectively catching fish. Polychlorinated biphenyls (PCBs), a group of organochlorides used in industrial and commercial applications, were still elevated in the Great Lakes, and suspected as one factor that might cause this defect. An alternative hypothesis that is consistent with field data, and has the added benefit of being reproduced in captive cormorants, is a deficiency of vitamin D in the diet.[21]

The story of one particular cormorant was used to draw attention to contamination in the Great Lakes. In 1988, Jim Ludwig visited Naubinway Island in northern Lake Michigan and discovered an approximately two-week-old chick with such an extreme cross-bill that he could not imagine how the parents had been able to feed it. Because debate over rules governing PCBs was ongoing, Ludwig decided to "adopt" the chick and used it as a poster child for the movement to clean up the lakes. Ludwig named the cormorant Cosmos, and she lived for thirteen months. Cosmos accompanied Ludwig on many educational presentations, made sixteen television appearances, and reached an international audience. Her presence reinforced Ludwig's oral testimony to legislative committees in Michigan, Wisconsin, and Washington, D.C, and greatly increased awareness about the use of toxic chemicals. During her short life, Cosmos became a potent symbol of the damaged Great Lakes ecosystem and was a powerful influence on decisions not to relax PCB regulations.[22]

Meanwhile, despite some continued exposure to PCBs and other chemicals, overall population growth continued at a phenomenal rate. By 1991, the Great Lakes cormorant population was estimated at more than 38,000 pairs. As the 1990s proceeded, the growth rate slowed significantly, but the population still more than doubled by 1997, reaching some 88,000 pairs. Finally, around 2000, the population peaked at approximately 115,000 pairs.[23]

There were also significant changes elsewhere in the interior. Breeding records for Illinois, Indiana, and Ohio showed a return after decades of absence or near absence; in Ohio, more than a century passed before cormorants returned as breeders. Regular breeding also resumed or increased in Iowa, Idaho, Nebraska, and North Dakota. Populations increased or developed in inland portions of New York, Minnesota, Wisconsin, and Vermont. At Minnesota's Lake of the Woods, changes paralleled those taking place in larger waters, and the cormorant population grew from just 100 pairs in 1976 to nearly 5,000 pairs in 1989. On the Ontario side, the period of population expansion was not captured, but the first large-scale survey, undertaken in 1983, estimated some 6,000 pairs were present, indicating substantial change had already occurred.[24]

In the southeastern portion of the continent, changes in wintering numbers mirrored those in the breeding grounds. Data from annual Christmas Bird Counts (CBC) indicate substantial increases from the late 1970s through the 1990s. Because that portion of the range draws birds from a nearly continent-wide breeding range, increasing concentrations of wintering birds there were particularly dramatic. By the mid-1980s, the CBCs indicated some

150,000 birds were present; the actual population size was likely substantially higher, since CBCs sample populations at only a limited number of locations and do not attempt to estimate actual sizes. For breeding birds in this region, twentieth-century declines and increases paralleled those in the interior and along the northern Atlantic Coast, and their timing suggests that cormorants in the Southeast were affected by some of the same factors influencing their populations in the North, especially contaminants and persecution. Similarly, the cormorant's recovery in the Southeast represented the return of breeding birds to states from which they had been absent for several decades, including Arkansas, Mississippi, and Tennessee, along with a northern expansion of the *floridanus* subspecies. The cormorant was also documented as a first-time breeder in South Carolina and Georgia, but in the former, recent breeding may represent recolonization after a long absence, since cormorants were abundant breeders in North Carolina and possibly the surrounding area during the colonial era.[25]

On the Pacific Coast, cormorants likewise began to recover, but not as dramatically as in the interior and eastern portions of the continent, and in several areas they did not regain their former numbers. In Alaska, a total of just 2,811 pairs were estimated, mostly at remote coastal sites, between 1970 and 1992. The western Aleutians had not been reclaimed, and numbers continued to represent what was likely a population diminished by introduced foxes and other animals. From British Columbia to Southern California, about 14,300 pairs were estimated along the coast between 1968 and 1992. During that time, the most dramatic increase occurred in the Columbia River estuary. But in the mid-1990s, human disturbance, predation by other birds, changes in feeding habitats, and, in some areas, persistent pesticide use continued to present conservation problems. While surveys in 2009 indicated the coastal population from British Columbia through California continued to increase to about 21,000 pairs, the change is mostly due to increases in the Columbia River estuary, where over 12,000 pairs occurred by 2009 and represented 60 percent of the coastal population. Simultaneously, declines occurred in British Columbia, northern Washington, and Southern California, attributed largely to disturbance, predation, and pollutants. In Mexico, monitoring efforts for seabirds are conducted in some regional areas, but recent survey data are limited, and it is not possible to assess the degree of recovery that may have occurred since the 1970s.[26]

Conservation Success Story or Environmental Demise?

The perception of how this second recovery was perceived is another fundamental thread in the cormorant's journey through time. Initially, changes were viewed as part of a conservation success story. But the large numbers of cormorants that suddenly appeared in places where the bird had long been diminished, gone, or perhaps never before present led many to question whether something other than conservation efforts led to the tremendous population changes. In several areas, the great rafts of cormorants fishing the waters, and the dense concentrations of birds standing erect and soldier-like on island shores, began to be regarded as unnatural phenomena. How, people wondered, had cormorant populations become so large so quickly? No single factor explains the cormorant's enormous population growth during this period. Rather, multiple influences arising from specific local conditions, combined with the cormorant's inherent reproductive capacity, appeared to be at work.

Undoubtedly, diminished human persecution and reduced pesticide contamination were the primary factors making recovery possible. Without those efforts, intact eggs and fledged chicks could not have occurred. But in several areas, additional changes may have enabled more cormorants to survive and raise larger numbers of chicks than otherwise would have resulted. On the breeding grounds, significant shifts in fish populations occurred at several locations as a direct or indirect result of human activities. In turn, altered fish populations may have increased food resources available to cormorants, leading to increased reproductive success. In the Great Lakes, the first half of the twentieth century brought pollution, extensive overharvesting by commercial fisheries, and the introduction of the sea lamprey and many other exotic species. Those pressures led to well-documented declines in large native predatory fish stocks between the 1940s and 1960s. In turn, small forage fish species attractive to cormorants, especially nonnative alewives and rainbow smelt, dramatically increased between the late 1950s and 1970s. Observations on the cormorant diet between 1979 and 1981 indicated that early in its recovery, the cormorant was taking full advantage of these highly abundant forage fish, particularly alewives, and achieving high reproductive success.[27]

Likewise, in the Prairie Provinces and along the northern Atlantic Coast, similar changes were documented during the same period. On Lake Winnipegosis, excessive commercial exploitation of fisheries resulted in a significantly changed fish community. Large predatory fish such as walleye declined, while forage fish like suckers and perch increased. Additionally, the forage-fish species tended to occur in large schools and in shallower areas of the lake, and were probably more vulnerable to cormorant predation. Comparisons between cormorant diet studies undertaken in the 1980s with those conducted in the 1940s indicated that the cormorant diet had changed, reflecting the abundance of forage species available. Similarly, at Lac la Biche, Alberta, decades of excessive commercial harvesting of walleye and pike led to the fishery's collapse in the early 1960s, and the fish community rapidly became dominated by small forage species such as perch, cisco, and shiners. In the estuary and gulf of the St. Lawrence, commercial overfishing during the twentieth century led to the collapse of the cod fishery and facilitated an increase in small fish species such as sandlance and capelin. Studies conducted in the region in the mid-1990s indicated that the cormorant diet consisted primarily of those schooling species, along with a few bottom-dwelling fishes, especially gunnels. While historical data on the cormorant diet in the region are limited, more recent data suggest that a change in the diet may have occurred over time, and fish-community changes have been linked to increases in double-crested cormorants and a variety of other colonial waterbird species in the area.[28]

On the wintering grounds, the creation of new food sources occurred during the same time that fish resources were changing on the breeding grounds, mainly in the form of aquaculture ponds and the proliferation of reservoirs. By the mid-1980s, thousands of acres of large, shallow ponds densely stocked with catfish and baitfish had been created in the southern United States. Additionally, thousands of acres of impounded waters containing a variety of suitable sport and rough fish species could be found in interior areas of the winter range where none had previously existed.[29]

While these changes supplied new food resources for cormorants, just how much each facilitated the species' recovery is not clear, because about the same time they became available, the cormorant was suddenly freed from the two best-established factors limiting reproductive success and population size—chemical contaminants and human persecution. As soon as those "brakes" were lifted, cormorant population growth resumed across much of its range, and not just in areas where new or enhanced feeding opportunities had been created. In this regard, data collected in two separate areas of the Great Lakes during the early to mid-1980s are particularly relevant. In the Apostle Islands, Lake Superior, a study conducted by University of Wisconsin researchers determined that cormorants existed on a diet consisting largely of sticklebacks, sculpin, and burbot during a period of significant population increase. The authors reported an inability to "identify any human factor related to the increase that might allow the National Park Service to interpret the increase as 'unnatural.'" Similarly, the cormorant diet was studied in Lakes Superior, Huron, and Michigan during the 1980s, and in Huron and Michigan, the primary prey of cormorants, alewives and smelt, experienced periods of significant decline. To compensate, cormorants shifted their diet to sticklebacks, suckers, and sculpins, and populations continued to grow.[30]

Likewise, while catfish ponds on the wintering grounds represent a superabundant food source, their importance to cormorant populations in general is overstated. For instance, on the breeding grounds, significant increases in cormorants had occurred by the late 1970s in several areas (for example, locations in the Prairie Provinces and the northeastern Atlantic), whereas cormorants did not develop a significant presence in the Mississippi Delta until the 1980s (see chapter 6). Additionally, the number of birds that use this region, while large from the human perspective, represents but a fraction of the continental population. In the early 1990s, the number of birds overwintering in the Mississippi Valley was probably equivalent to no more than 5 percent of just the interior population, with the vast majority of birds overwintering coastally and elsewhere and never visiting a catfish pond. Additionally, large increases in the Delta were mirrored by large increases in other areas of the wintering range. Based on CBC data obtained through 1992, the largest inland numbers for Mississippi occurred at Natchez, when more than 20,000 birds were counted. This CBC location is well south of the Delta, in the same area where Audubon, in December 1820, observed *millions* of cormorants flying southwest in long lines for several hours, indicating a historical precedent for large numbers in the region.[31]

Regarding the perception of unnatural population size, whether population growth would have proceeded more slowly and what size populations would have reached in the absence of these new resources will never be known. But a few important points relative to these questions should be emphasized. First, historical records indicate that cormorants existed in very large numbers across much of the continent when they were initially encountered by European settlers. Furthermore, those records span hundreds of years and locations, indicating that cormorants had been an abundant and widespread species for a substantial time. Presumably, presettlement populations represented

abundance attained under more natural conditions, that is, before human activities such as overharvesting by fisheries, the introduction of exotic species,
and the alteration of habitat could inflate or diminish cormorant numbers.
Moreover, cormorants had once existed in large numbers in many areas where
new fish resources developed. Thus, there is historical precedent for large cormorant populations, not just generally but at specific locations where concerns
developed.

Second, several aspects of the cormorant's natural history and behavior allow it to quickly respond to changing environmental conditions. As fish
populations move about in the waters to spawn, feed, or respond to environmental changes, cormorants are able to quickly alter their foraging strategy
to take advantage of new fishing opportunities. What can be stated with some
certainty is that the cormorant's extremely adaptable feeding behavior grants
it a certain resiliency during times when preferred or easy prey are less abundant. The birds' flexibility in locating nests also confers important advantages.
Furthermore, cormorants begin breeding relatively early compared to other
seabirds and waterbirds, with most birds initiating nests by the age of three and
some perhaps as early as the age of one or two. The number of eggs laid and
the typical rates of reproductive success experienced are also relatively high,
and if a clutch is lost early in the nesting season, the cormorant will frequently
re-lay and may manage to produce some chicks. Those features, according to
the evolutionary biologist Doug Causey, predispose this species to success but
also to conflicts with humans.

Finally, an important way of looking at why populations became so large
is to revisit the question of what limits population size. As predators that reside
high up in the food chain, cormorants appear to be most significantly influenced by the abundance and availability of prey fish. Cormorants are thus an
important ecological indicator; if their numbers are "out of balance" or "unnatural," so too is the ecosystem in which they live. By the same token, abundant cormorant populations indicate an aquatic environment healthy enough
to support large concentrations of birds and fish. How cormorant population
growth is viewed depends largely on whether the fish species supporting cormorant numbers are those preferred by humans. Where this is the case, cormorants are often perceived as direct competitors and as a threat to human
interests. But even where cormorants consume forage or rough fish, they are
often still perceived to threaten human interests, equivalent to weeds thriving
in an altered community. In the latter case, cormorants are seen as symptom-

atic of an environment's "degraded" quality and as preventing the recovery of fish species that humans are interested in catching. Good examples of this scenario are the conflicts that developed over fish at Lac La Biche and Lake Winnipegosis.

This double-sided aspect of just what large numbers of cormorants represent is a fundamental component of the human-cormorant conflict, and one in which the cormorant seemingly can't win no matter which side of the coin is up. And so, as events unfolded during the most recent phase of the cormorant's recovery, a by now familiar pattern characterizing human-cormorant interactions again emerged: as cormorant populations increased in size and range, so too did conflicts between humans and birds.

PART III

The Economic and Political Landscape of the Cormorant, 1965 to the Present

6

Fish Ponds and Reservoirs

The Context for Conflict on the Wintering Grounds

IN THE DELTA region of Mississippi, many factors have converged to create a perfect storm that swirls around the production of the channel catfish, *Ictalurus punctatus*, and its consumption by the double-crested cormorant. Here, the war over catfish has steered the cormorant's more recent story and given it a distinctly US orientation. It has also done more to establish the cormorant as a continental problem than any other issue during the past thirty years.

Located in northwestern Mississippi, the Delta region, commonly referred to as the Mississippi Delta, is a 6,200-square-mile alluvial plain that encompasses the floodplain of the Mississippi River and some of its tributaries— the Yazoo, Sunflower, and Tallahatchie Rivers. The region is well suited for large-scale mechanized agriculture, and much of the Delta has been drained for this purpose. Mostly flat topography and impervious soils are two of its most attractive agricultural features, and the region was recognized as an ideal environment for catfish farming nearly fifty years ago. The first commercial catfish ponds were constructed in 1965 in Sharkey County, in the eastern portion of the region, when cotton prices dropped and farmers pursued other uses for their lands. Although Arkansas is the birthplace of the commercial catfish industry, where at least two farms were selling catfish in the late 1950s, it was

Channel catfish, at the eye of the storm

production in the Delta region that transformed the industry into an economic giant. By 1976, there were some 54,000 acres of catfish ponds in production in the United States, about half of them in Mississippi. Humphreys County, just northeast of Sharkey, produced more catfish than any other county in the country, and was befittingly proclaimed the "Catfish Capital of the World."[1]

This title is no small claim to fame, and competition for it gives some idea of the iconic status the catfish holds in the Southeast. The title had actually been claimed earlier in Savannah, Tennessee, located on the Tennessee River, a world-class fishing destination for those pursuing jumbo blue, channel, and flathead catfish. In the 1950s, the city added the title to its postmark in an effort to distinguish its mail from that heading to Savannah, Georgia. The city hosts a yearly "National Catfish Derby," a six-week summer fishing competition, and to this day calls itself the "Catfish Capital of the World."

Some twenty years later, the town of Des Allemands, Louisiana, situated along the Bayou Des Allemands, also claimed the title. Here, the catfish is king of the local fishing industry, and in 1975, the town held its first Louisiana Catfish Festival in an effort to stimulate the local economy. To help promote the event, Louisiana's then governor, Edwin Edwards, signed a proclamation declaring Des Allemands the Catfish Capital of the World. In the following year, acquisition of the title by Humphreys County, Mississippi, put three separate and global catfish capitals on the map. In an apparent effort to one-up the other two locations and further distinguish Des Allemands, the Louisiana legislature passed a resolution in 1980 renaming the town the "Catfish Capital of the Universe." But by then, the Mississippi Delta had become the most concentrated area of aquaculture production in the United States, and Humphreys County went on to develop more catfish acreage than not only any other county but any other state in the country, greatly strengthening the county's claim to the title.

During the 1980s, the catfish industry grew tremendously; by 1993, it was producing an estimated 225,000 tons of catfish, valued at $353 million, and accounting for about half of the value of US aquaculture. Mississippi was churning out 75–80 percent of this total, and most catfish were produced in the Delta. Meanwhile, Alabama, Arkansas, and Louisiana too were becoming significant catfish production states. Industry growth continued, and peaked between 2002 and 2003 at about 198,000 acres of ponds in production and over 300,000 tons of fish processed, again with most of the production in the Delta.[2]

In the late 1990s, US producers began experiencing stiff competition from Vietnam and China, which were exporting catfish to the United States. Additionally, competition from imported tilapia, and within the seafood marketplace in general, intensified. In 2005, foreign competition jumped to a new level. At about the same time, higher costs for grain and energy led to dramatic increases in operational costs, while the weak economy made it more difficult for producers to obtain credit and stay in business. These global economic forces have taken a significant toll on the industry. By 2011, the amount of catfish raised and processed in the United States had decreased by 49 percent from its peak in 2003, while foreign imports had increased dramatically, accounting for 74 percent of all US sales of frozen filets. Additionally, since 2002 the amount of water surface area in production in the four major catfish states has sharply declined, ranging from a loss of 29 percent in Alabama to 93 percent in Louisiana, where the industry has all but collapsed. Many producers in these states have converted their ponds to corn and soybean fields. But while domestic catfish production is clearly struggling and faces an uncertain future, it is still an economically significant industry. In 2011, domestic catfish sales equaled $423 million, and at the start of 2012, 89,000 acres of surface water were in production, most of it in the Delta.[3]

While the Delta region's unique physical attributes make it ideal for many agricultural practices, it is crisscrossed by migratory pathways used by birds traveling to and from wintering and breeding grounds. Its western portion lies at the confluence of the Mississippi and Arkansas Rivers, two drainages that bring vast numbers of birds to and from the region annually. The Mississippi River channels northward and southward movement along the length of the country by millions of birds, many of which pass the Delta region on their way to and from breeding and wintering areas. The river's many tributaries likewise provide well-defined migration pathways to and from the area, and the Arkansas River facilitates movement of birds in and out of the region from the west. In addition, the region retains more than 10 percent of its original wetlands, and its remaining cypress swamps, oxbow lakes, and bayous provide ideal wintering and breeding habitat for cormorants and other fish-eating birds. These features made the inevitable discovery of the region's abundant catfish ponds by fish-eating birds only a matter of time.[4]

But location was not the only reason conflicts developed between humans and fish-eating birds; the design of the ponds themselves and the farms' production systems all but guaranteed there would be problems with bird predation. The first ponds were built when cormorants had no serious presence in the region, and facilities were not constructed to exclude or discourage cormorants or other birds from using them. Rather, they were designed to maximize fish production, and their physical dimensions were described as "the secret to the catfish farmer's success." Ponds are typically rectangular and large, especially in the Delta. In the mid-1980s, most ponds there were about 20 acres in size and shallow, three to six feet deep. In 2010, large, shallow, rectangular ponds remained the standard, with an average pond size in the Delta of almost 15 acres. The scale of most operations is large, with the average farm having more than 250 acres under production. Narrow levees are built alongside the ponds so that trucks and tractors can move about to maintain and harvest fish. At harvest time, nets as wide as 100 yards are strung between tractors on both sides of the pond and dragged along its length to scoop up the fish.[5]

Three types of ponds are used: brood fish ponds, where breeding animals live and from which eggs are harvested; fry and fingerling ponds, where young fish grow until they reach "stocker" size; and food fish or grow-out ponds, where larger fish grow until they reach harvestable size. Incredibly dense stocking rates are employed, with anywhere from 4,000 to 100,000 fish per acre being stocked, depending on the pond type. About 90 percent of all ponds are used for food fish, and are stocked at densities between 4,000 and 10,000 fish per acre. Most farmers use a multiple-batch production and cropping system in which fish of different sizes and ages are present in a food fish pond at any given time. With this system, there is no need to drain the pond at harvest; rather, fish farmers use seine nets with a mesh size that selectively captures fish large enough to be harvested. The smaller fish are left in the pond to grow, and the fish that have been removed are immediately replaced with "stocker size" fish.

Less commonly used is the original single-batch cropping system, in which fish hatched during the same year are all grown together. When the fish are big enough to harvest, the pond is drained and the fish retrieved. The single-batch system has several advantages over the multi-batch system, including the production of fish that are more uniform in size; more accurate inventory records, since fish populations can be "zeroed out" after each period of harvest; and feed conversion efficiencies that tend to be greater. Nevertheless, the multi-batch system is preferred because it creates a stable flow of fish to processors and cash to producers, and farmers can go for years without drawing down ponds. But from the perspective of bird predation, this system provides a greater distribution of fish in size ranges that are vulnerable to predators all year long.[6]

For fish-eating birds, the shallow depths, dense stocking rates, and lack of cover in ponds, along with the constant availability of some fish in the appropriate size range, make fishing in the ponds akin to fishing in a fish bowl. While the pond dimensions and catfish production systems used maximize profits, they make covering the ponds with nets to exclude fish-eating birds an enormous investment. Additionally, the levees between the ponds are too narrow to accommodate both extensive net structures and tractors moving back and forth. Low profit margins and the high cost of modifying farms, as well as the impracticality of covering ponds harvested as frequently as once a week, have been identified as factors preventing producers from bird-proofing their facilities.

Thus, the table was lavishly set and the guests arrived. Cormorants, anhingas, pelicans, herons, and egrets all keyed in on the vast and expanding resource and became frequent visitors to catfish farms. Many of these species were typically present in the Delta region year-round, such as great blue herons, or during a portion of the year, such as great egrets. Cormorants historically had a smaller presence in the region, but their foraging behaviors are perfectly suited to the catfish ponds. In the past, cormorants had typically migrated down the Mississippi River past the Delta region to wintering areas farther south along the Gulf Coast. Some range maps suggest they wintered as far north as the Delta region, but populations of birds known to occur historically along the lower portion of the river were believed to have been very small. But as cormorant numbers recovered across the range, migrating birds encountering this cornucopia began stopping for longer periods of time or spending the entire winter in the Delta. In 1982, cormorants were detected for the first time during Christmas Bird Counts in Washington County, Mississippi, the CBC site closest to the region's concentration of catfish farms. Numbers grew rapidly, and fish farmers began to voice complaints regarding losses of catfish. Cormorant numbers continued to increase, and by 1990 some 30,000 were estimated to occur in the Delta region. Diet studies of birds shot at ponds and nearby night roosts in the early 1990s indicated that catfish were an important part of the cormorant's diet, along with gizzard shad, a common forage fish of the Mississippi River that is known to invade catfish ponds. A later study examining body masses suggested that cormorants exploiting catfish farms may have improved body conditions and increased winter survival rates, allowing these adults to return to breeding grounds and contribute to population growth.[7]

Other Cultured Fish and Animals

Upon return to the wintering grounds, the cormorant not only encountered rich, easy dining in the catfish ponds of the Delta and elsewhere, but also in ponds stocked with baitfish and crayfish. Cultured baitfish commonly include minnows, shiners, and common goldfish, and this industry was also expanding in the Southeast as cormorants were recovering. By the mid-1990s, the baitfish industry had a farm gate value exceeding $40 million, with 90 percent of production occurring in Louisiana and Arkansas. And like catfish, baitfish are cultured in ponds attractive to and easily accessed by fish-eating birds. But the main predators at these ponds are herons and egrets rather than cormorants, although cormorants also consume some of these fish.[8]

The crayfish industry, too, is centered in the region, most production being located in Louisiana but with some in Mississippi and Arkansas. Like other types of aquaculture, the industry grew as cormorant numbers were increasing. Between 1980 and 1991, production tripled in Louisiana, with a farm gate value exceeding $31 million. Crayfish are raised in large ponds and, like

cultured fish, are vulnerable to bird predation. The most significant avian predators include ibises, night-herons, and egrets, but cormorants also eat crayfish,
tip over traps trying to get them, and eat the fish bait set out for the crayfish.[9]

Because of escalating conflicts with fish farmers, the Arkansas legislature passed Senate Bill 345 in 1993. It declared the cormorant a "nuisance"
to the fish-farming industry and encouraged elimination of the cormorant's
protection under the Migratory Bird Treaty Act. This bill resulted in Act
575, whereby the cormorant was officially declared a nuisance in the state of
Arkansas.

Reservoirs

As the catfish industry was expanding in the Southeast, so too were the number and size of reservoirs being created by the US Army Corps of Engineers
and others. These impoundments, or "lakes," as they are often called, were
created for multiple uses, including water supply, flood control, hydroelectric
power, recreation, and fish and wildlife, or some combination thereof. They
vary greatly in size, and can be as small as ten acres to as large as hundreds of
thousands of acres in area. Their storage capacity, another indicator of size,
is described in terms of acre-feet, which refers to the amount of water necessary to cover one acre with water one foot deep. Rapid growth in reservoir
construction across the United States occurred between about 1850 and 1960.
By 1963 there were 1,562 bodies of water described as reservoirs or controlled
natural lakes that were either completed or under construction, and that had a
usable water capacity of 5,000 acre-feet or more. Their combined usable storage equaled about 360 million acre-feet, and their combined surface area approximately 15 million acres.[10]

Construction of reservoirs has continued, and for cormorants and many
other fish-eating birds, these waters are attractive feeding areas in the winter
or during migration. In the Southwest, reservoirs and impoundments are especially abundant. In Texas there are more than 6,700 reservoirs with a surface
area of 10 acres or larger, and in 2011 there were 187 with a capacity of 5,000
acre-feet or more. For comparison, there were only 117 reservoirs with that
capacity in 1966, and just 4 in 1913. Likewise, thousands of reservoirs have
been created in Oklahoma, where construction swelled between the 1930s and
1970s. Currently, this state has 61 reservoirs with a capacity of 11,000 acre-feet

or more, and over 100 that are considered major (5,000 acre-feet or more in capacity). For comparison, in 1966 there were only 36 reservoirs with a capacity
of 5,000 acre-feet or more.[11]

In these interior areas, reservoirs have also become highly popular sportfishing destinations. To enhance fishing opportunities for anglers, many are
stocked by natural resource agencies with sport-fish species that are often nonnative or unable to successfully reproduce and maintain abundant populations.
As cormorants recovered and encountered these waters, conflicts with anglers
began to occur. By 1991, complaints from anglers led to the introduction of
Senate Bill 362 in Oklahoma, which declared the double-crested cormorant
a "nuisance." The bill was signed into law, making the cormorant's nuisance
status official in the state and sending a clear message to the USFWS that expressed "the feeling of many Oklahoma sportsmen." Cormorant fishing at reservoirs was also part of the impetus for the nuisance status assigned to the bird
in Arkansas, and for the pest status that emerged in Texas in the late 1990s. In
the latter state, cormorants are regularly seen at nearly all large public reservoirs and during most fishing trips. In a 2007 *San Angelo Standard-Times*
article on cormorant control in Texas, Terry Maxwell, a biology professor, put
the issue in the following perspective: "It is ironic in the extreme that this issue
is visiting drought-plagued west-central Texas—a land that only a hundred
years ago was without attraction to cormorants. But we then built dams, impounded lakes and stocked them with a fish buffet. What now are we going to
do with the uninvited guests?" Although the comments were specific to Texas,
they essentially sum up this particular problem for many parts of the southern
United States.[12]

7

Animal Damage Control and
the First Standing Depredation Order
for Cormorants

IN THE UNITED STATES, protection under the Migratory Bird Treaty Act (MBTA) makes taking, killing, or even possessing migratory birds, along with their nests or eggs, illegal. Nonetheless, there are situations in which protected birds can be legally killed, or "taken." The *Code of Federal Regulations*, which describes the rules governing the MBTA, identifies circumstances in which protected migratory birds can be so managed. The circumstances include the control of "depredating" birds—those birds considered harmful to human interests because of their consumption or use of agricultural crops or aquacultural products.

To manage nongame migratory birds involved in depredation issues, the US Fish and Wildlife Service has developed two procedures. First, the service (and only the service) issues depredation permits, which have specified conditions and durations. For most species, it is through such federally issued permits that birds are taken. For a much smaller number of species, including blackbirds, crows, and other corvids, various crowned sparrows, house finches, and purple gallinules, the service has established standing depredation orders. These orders waive the requirement for a federally issued permit, but define specific conditions under which birds can be taken.

Landing on the water

In 1986, the USFWS took its first significant action to address the burgeoning conflict between fish farmers and cormorants in the southern United States, issuing depredation permits to fish producers, which allowed them to shoot cormorants at their ponds. To obtain a permit, fish farmers were required to demonstrate that they were experiencing economic impacts from cormorant predation and that they had tried nonlethal alternatives without success. In this regard, the year 1986 marks a turning point in the cormorant's history as the beginning of management on the wintering grounds.

A far more significant event occurred in the same year when the US Congress passed legislation that dwarfed the importance of this initial management measure. The legislation transferred the federal Animal Damage Control (ADC) program from the USFWS, part of the US Department of Interior, to the Animal and Plant Health Inspection Service (APHIS), part of the US Department of Agriculture (USDA). The mission statements of these agencies, posted on their websites, provide an inkling of the philosophical shift this move entailed. The mission of APHIS is to "protect and promote U.S. agricultural health, regulate genetically engineered organisms, administer the Animal Welfare Act and carry out wildlife damage management activities," while that of the USFWS is "to conserve, protect and enhance fish, wildlife, and plants and their habitats for the continuing benefit of the American people." The transfer changed the infrastructure not only for cormorant management but also

more broadly for all wildlife deemed a nuisance. To understand the significance the transfer has had for cormorants in particular, we must take a brief
historical tour of the development of the Animal Damage Control program in
the United States.

The Evolution of Animal Damage Control / Wildlife Services

Animal Damage Control is a federal program now known as Wildlife Services.
It traces its origins to an 1885 congressional appropriation of $5,000 to orga-
nize the Section of Economic Ornithology as part of the Entomology Division
of the USDA.[1] The section, which grew out of the American Ornithologists'
Union's (AOU) Committee on Bird Protection (established in 1884), began
collecting data on the agricultural benefits of insect-eating birds and birds of
prey. In 1886, a separate Division of Economic Ornithology and Mammalogy
was created, mainly to educate farmers about birds and mammals affecting
their interests in order to avoid destruction of the many useful and nonharmful
species that were being persecuted at the time.[2] In 1890, the division became
simply the Division of Ornithology and Mammalogy, and while it continued
to advise farmers, its focus expanded toward surveys of animal distributions
and habits. To reflect this broader emphasis, the name was changed in 1896
to the Division of Biological Survey, and in 1905 was slightly modified to the
Bureau of Biological Survey. That name was maintained for the next twenty
years, but underwent periodic changes to reflect the bureau's evolving identity.
Within the bureau, names for the animal-damage control program likewise
went through several iterations. Each attempted to define the program's purpose, reflect its approach toward human-wildlife interactions, and especially
important, create a particular image.

As the Biological Survey was developing, it embarked on multiple ani-
mal-control projects, and in 1905 began working with the US Forest Service to
develop ways to control wolves and coyotes. Ranchers were grazing livestock
on national forest lands, and the Forest Service had begun to charge fees for
the privilege. But since cattle and sheep were being lost to predators, ranchers objected to paying fees. In 1915, under pressure from ranchers, Congress
made sizable appropriations for the control of predators and called for direct
participation by the Biological Survey. The bureau extensively trapped, shot,
and poisoned animals, including the use of massive nonselective poisoning
campaigns for a variety of species. By the 1920s, untold numbers of wolves,

coyotes, and other predators, along with vast numbers of prairie dogs, rabbits, rodents, bats, and birds, were being killed. Those efforts eventually led to such infamous results as the extirpation of the gray wolf from almost all its range in the lower forty-eight states, a project that had been in the making long before the federal government stepped in, but whose involvement effectively sealed the animal's fate. The coyote also became a defining animal for the agency, but being more resilient than the wolf, it withstood some of the greatest animal-damage control efforts in the program's history. Between 1916 and the mid-1980s, more than four million coyotes were shot, poisoned, trapped, gassed, and otherwise destroyed through animal-damage control efforts alone. The program continues to kill an average of 82,000 coyotes a year, according to APHIS data for the period between 2000 and 2011.[3]

As techniques to manage wild animals developed, the Eradication Methods Laboratory was established in Albuquerque, New Mexico; experimentation with toxicants was the lab's primary focus. In 1921, the lab moved to Denver, Colorado, and eventually became the National Wildlife Research Center, now a state-of-the-art facility for developing solutions to human-wildlife conflicts within the APHIS–Wildlife Services division. In 1924, the Biological Survey named its animal-damage control program the Division of Predatory Animal and Rodent Control (PARC). The program continued to expand and grow, but its activities were not uniformly accepted. In 1931, the American Society of Mammalogists adopted a resolution in which it publicly challenged and condemned the Biological Survey: "The executive officers of the Survey have brought about the crisis confronting our wildlife *and are more directly responsible for it than any other agency*. . . . The issue at stake is not for us to prove that the Survey is wrong, but for the Survey to prove that it is right. The Society . . . by every dictum of logic and common sense can call upon the Survey to *show full and adequate cause for becoming the most destructive organized agency that has ever menaced so many species of our native fauna*." Although Congress held hearings on PARC, in the end the only result was the Animal Damage Control Act of 1931, which authorized direct and cooperative control programs by the division. The act specifically directed the secretary of agriculture to "determine, demonstrate, and promulgate the best methods of eradication, suppression, or bringing under control on national forests and other areas of public domain as well as on State Territory or privately owned lands [predators and rodents] and other animals injurious to agriculture . . . and to conduct campaigns for the destruction or control of such animals."[4]

Efforts to rein in the Biological Survey continued, and the Society of Mammalogists, along with numerous scientists from natural history museums and research institutions, joined forces with the outspoken environmental activist Rosalie Edge. Edge engaged the public on the issue by, among other means, issuing the pamphlet *The United States Bureau of Destruction and Extermination: The Misnamed and Perverted Biological Survey* (1934). Edge and her supporters charged that PARC's methods were "reckless, cruel and indiscriminate, and that its purposes and activities are to capitalize, increase, inflame and use for its own advantage the prejudices and enmity . . . that people feel towards so many of our wild neighbors."[5] In her pamphlet, Edge presented a petition calling for the extensive program of destruction to be abandoned immediately. The petition was signed by many prominent scientists from universities and natural history museums across the country. But despite this avalanche of scientific opposition, PARC continued poisoning, trapping, shooting, and otherwise destroying wild animals.

In 1939, the USDA's Bureau of Biological Survey and the Department of Commerce's Bureau of Fisheries were both transferred to the Department of the Interior, and were combined the next year. That move brought PARC under the authority of the federal department responsible for protecting America's natural resources. Further reorganization in the 1950s led to the creation of the USFWS, the agency in which PARC would be housed. Initially, the move did little to change animal-damage control policy, and the status quo was maintained until the 1960s, when the environmental movement got under way. By 1963, public appreciation for wildlife and awareness of toxic chemicals was becoming widespread, and opposition to animal-damage control activities was growing. Public concern led the secretary of the interior, Stewart Udall, to convene an advisory board on wildlife management to review predator control in the United States, especially the activities of PARC.

Headed by Starker Leopold, the eldest son of Aldo Leopold, the board produced the highly critical "Leopold Report." The review concluded that federal responsibility for minimizing wildlife damage was properly assigned to the USFWS but, in charges reminiscent of those brought by Rosalie Edge and others in the 1930s, deemed PARC's actions nonselective and excessive. The board asserted that PARC killed far more animals than was required, had "become an end in itself," and had "developed into a semi autonomous bureaucracy whose function in many localities bears scant relationship to real need and less still to scientific management." The board made six recommendations

to address these abuses: appoint an advisory board reflective of multiple interests; reassess the goals, function, and purpose of the program, in light of
changing public attitudes toward wildlife; revise predator and rodent control
guidelines and justify the need for control with quantitative statistics on the
true extent of damage; greatly amplify the research program, especially in
regard to developing nonlethal methods and controls that would reduce unnecessary killing; change the name of PARC to reflect the basic changes in
policy and philosophy recommended by the advisory board, specifically to one
that reflects the "much broader responsibility for management of animal life
in ways other than killing"; and legally regulate the use and distribution of
poisons, especially compound 1080, which targeted canids but was a general
toxicant that killed relatively indiscriminately, was inhumane, and had broad
effects.[6]

Despite these recommendations, the only visible change made to the
program was to its name. After the "Leopold Report" was published, PARC
was rebranded as the more politically correct Division of Wildlife Services,
and other changes of a primarily administrative nature were made. Meanwhile,
many of the division's excessive practices continued. In 1971, several environmental and animal welfare groups, including Defenders of Wildlife, the
National Wildlife Federation, the Sierra Club, and the Humane Society of the
United States, brought lawsuits against the Department of the Interior. In response, the department convened another advisory board to review the animal-damage control program. In 1972, it produced the "Cain Report," which
reiterated many of the Leopold Report's recommendations, but also went beyond them. For example, it suggested that landowners be taught to control
predators themselves, rather than to rely on control by federal agents.[7]

The Cain Report led President Nixon to sign Executive Order 11643,
which banned the use of poisons for predator control by a federal program
and on federal lands. Additionally, the EPA canceled the registrations for compound 1080, strychnine, and sodium cyanide, poisons of choice for animal-
damage control. Subsequently, the Division of Wildlife Services began to rely
on aerial hunting and increased trapping efforts. A further effort to modify the
animal-damage control program, the Animal Damage Control Act of 1972,
was drafted by President Nixon and passed the House but failed to come to a
vote in the Senate.

In 1975, more restructuring occurred within the program. The Division
of Wildlife Services was dissolved, its Enhancement and Pesticide Branches

were moved to another USFWS division, and the branch of animal-damage control was reduced to an office in Washington, DC, and became known as the Office of Animal Damage Control. But President Ford made amendments to Executive Order 11643, allowing the operational use of sodium cyanide in the M-44, a device planted in the ground that appears similar to a survey marker and lures predators with scent or a small piece of bait. When the animal investigates the device, a spring shoots a dose of sodium cyanide into its face.

Environmentalists continued to call for additional changes to the ADC program, and in 1978, President Carter's interior secretary, Cecil Andrus, appointed yet another committee to review the program. In the third report to be highly critical of the program, the committee reportedly stated that it "found insufficient documentation to justify the program's existence." As a result, the USFWS prepared an environmental impact statement on predator control in the West. The combined reports led Secretary Andrus to issue a policy in 1979 that halted the practice of den hunting predators (in which predators were tracked to their dens and the pups inside were killed) and emphasized nonlethal predator control. In addition, Andrus banned further research on compound 1080, which the USFWS had continued to study in case the earlier ban were to be lifted.[8]

These policy changes elicited strong opposition from ranchers and the agricultural community, which had the support of not only ADC but also western senators and representatives. The Department of the Interior was accused of not respecting the needs of the livestock industry. In 1980, Congress requested that the president consider transferring ADC back to the USDA, but he declined. Toward the end of the year, Ronald Reagan was elected president, and most of the important controls on animal-damage management began to evaporate. In 1981, the Andrus policy on animal-damage control and the ban on den hunting were rescinded by Secretary of the Interior James Watt. In 1982, President Reagan signed Executive Order 12342, revoking President Nixon's Executive Order 11643; compound 1080 was reinstated for use in "livestock protection collars," poison-filled bladders strapped around the heads of sheep and goats. In 1985, President Reagan signed an executive order revoking President Nixon's environmental safeguards on animal-damage control. Concurrently, twenty senators signed a letter again requesting that the president move ADC back to the USDA, which happened in 1986, when the program was made an organizational unit of APHIS.[9]

With that transfer, the way in which wildlife damage problems were addressed in America set sail on a new course. Congress immediately provided a major boost in ADC's funding, establishing a $20 million budget at the time of the transfer. Shortly thereafter, the secretary of agriculture established an advisory committee to make recommendations on policies and program issues regarding wildlife damage control, called the National Animal Damage Control Advisory Committee (NADCAC).[10]

But unlike the earlier committees established by the secretary of interior, NADCAC was very supportive of ADC, and during its initial meetings identified several critical concerns for the program. Given ADC's inability to meet the increasing need for its services, the committee recommended that ADC's budget be increased substantially. By 1990, the program's proposed budget had been raised by some 54 percent, with a request for $30 million to accomplish its objectives. Another critical issue was cooperation between the USFWS and ADC in resolving problems related to migratory birds. ADC had responsibility for managing damage problems associated with migratory bird conflicts, but the USFWS retained regulatory authority over migratory birds. This distribution of responsibility and regulatory authority set the stage for conflict between the two agencies; ADC viewed the split authority as an impediment to its ability to accomplish its objectives. Almost immediately, ADC reported that it had "encountered some obstacles" as it attempted to address migratory bird problems, because it lacked the management authority held by the USFWS. The advisory committee recommended that "the Director of FWS and the Administrator of APHIS get together to work toward resolving some of the regulatory obstacles to dealing with these problems."[11]

Although the problem of shared responsibility seemed simple enough on the surface, in reality a far more complicated issue was developing. ADC'S identification of management authority as an obstacle to its work provides the first inkling that the program saw itself as an entity with its own direction. Given the USDA's focus on protecting resources from wildlife, ADC would require greater latitude and autonomy to accomplish its tasks. Furthermore, the program would not just provide on-the-ground services to implement cormorant management, but would ultimately evolve into one with a major role in developing management policy. To facilitate ADC's immediate responsibilities, the USFWS issued depredation permits for migratory birds to ADC in every state, authorizing the unlimited killing of nongame migratory birds for the purpose of resolving damage problems as ADC saw necessary.[12]

Wildlife Services Focuses on Cormorants

On return to the USDA in 1986, ADC immediately took the lead in addressing the cormorant-aquaculture conflict. The growing economic importance of the catfish industry escalated the problem of fish-eating-bird depredation to a major focus area for the program. Responsibilities included monitoring cormorant numbers in the Delta region, documenting economic impacts to the catfish industry from cormorant predation, and developing and implementing depredation-control technologies. Funding for these activities continues to be provided through annual state and federal allocations, and from those with a vested interest in the program, including fish producers, private individuals, and businesses.

In 1988, the National Wildlife Research Center established a research station in Starkville, Mississippi, specifically to address bird depredation, especially by cormorants, on catfish. One of the station's first projects was to survey catfish farmers of the Delta region regarding the methods they used to repel avian fish predators. Farmers reported harassing, shooting, and scaring birds through various means. ADC's survey also obtained the first estimates of potential economic losses to farmers in the region from cormorant predation. Using rough calculations based on the approximate number of cormorants in the region, and assuming that cormorants had a diet consisting entirely of catfish, ADC estimated that cormorants consumed $3.3 million dollars worth of catfish in 1988 alone. Including costs to repel fish-eating birds, which did not single out efforts used for cormorants, estimated expenses to the catfish industry because of birds totaled $5.4 million a year, about 3 percent of total sales.[13]

It is important to point out that producers may provide reasonable estimates of the amount of money they spend on bird control at their farms, but reliable financial estimates of fish lost to a particular predator, as calculated either by producers or scientists, have thus far remained elusive. As a result, all estimates of losses should be carefully considered to evaluate how realistic they may be. Those based simply on observations of birds consuming fish at individual ponds and then extrapolated to larger scales are especially subjective and limited in applicability. Further, reliance on untested assumptions, such as the claim that the cormorant diet consists entirely of catfish, makes estimates relying on such assumptions all the more questionable. The science used to assess cormorant impacts on aquaculture is discussed in later chapters.

While ADC was expanding its role in the research and management of fish-eating birds, the USFWS was being heavily criticized by the aquaculture industry for not supporting farmers in their efforts to deal with bird-caused losses. Although the service was issuing depredation permits, farmers considered the process slow and inefficient. In response, the USFWS in 1990 issued Director's Order No. 27, "Issuance of Permits to Kill Depredating Migratory Birds at Fish Cultural Facilities." The order granted permits to kill a variety of species and sped up the process of obtaining a permit to within seven days of a request. Under the order, thousands of birds were killed; nevertheless, the extent to which the landscape had been transformed by southern aquaculture kept a steady stream coming. As a result, from Louisiana to Alabama there were conflicts with numerous species of fish-eating birds, from kingfishers to pelicans. Those surrounding catfish and cormorants in the Delta region, however, continued to dominate the show, and cormorants represented the majority of birds killed at aquaculture facilities across the nation. Between 1993 and 1994, 2,261 people held depredation permits, and some 8,200 cormorants a year were shot. Most of these birds were killed in the Southeast, where the number of permits issued to aquaculturalists quadrupled between 1989 and 1996.[14]

Concurrently, ADC was investigating and reviewing the effectiveness of various nonlethal measures. Bird-frightening programs, including pyrotechnics, live ammunition, distress calls, electronic noises, scarecrows, and propane exploders, were used in various combinations and with varying degrees of success. Exclusion techniques were reviewed, and a few tried, but most were considered to have limited use or to be impractical because of pond design and harvesting techniques. But in the winter of 1988–1989, an effort to disperse cormorants from night roosts by using pyrotechnics and other disturbances proved highly effective in reducing the number of cormorants at nearby catfish farms. As a result, ADC launched a night-roost dispersal program in 1993 throughout the Delta region in conjunction with catfish farmers. The effort reduced damage from cormorants locally and was considered to have diminished the numbers of birds that otherwise would have been killed in the area.[15]

To refine economic estimates of cormorant-caused losses, ADC undertook several additional studies in the late 1980s and early 1990s to better evaluate the impact of cormorants in the Delta region. Data collected on cormorant population status, the use of winter roosts, and daily activity budgets and diet were incorporated into a bioenergetics modeling effort to estimate crop loss.

The results were published in 1995 in a special volume of the scientific journal *Waterbirds* that grew out of a cormorant symposium convened in Oxford, Mississippi, in 1992. Jim Glahn, a scientist from the National Wildlife Research Center in Starkville, was a principal investigator in many of these studies. The metric tons of catfish consumed by cormorants during the winters of 1989–1990 and 1990–1991, as estimated by the bioenergetics model, equaled approximately 18–20 million catfish per year. Most catfish consumed were large enough to be stocked in grow-out ponds, and it was assumed these fish would have been harvested had cormorants not eaten them. Therefore, the replacement cost of these fish was estimated and equaled about $2 million a year. The extent of loss at the time of catfish harvest, when fish are larger and many times more valuable than fingerlings, remained unknown. Nevertheless, the estimate was considered to provide the best estimate of cormorant impacts to date.[16]

The millions of fish eaten and the presumed income lost from cormorant predation represent large quantities. But some context for these estimates helps put industry impacts from cormorant predation into a broader perspective. The number of fish consumed, while large, corresponded to just 4 percent of the standing crop of catfish each year, indicating that cormorants caused relatively small losses to the catfish industry overall. Likewise, an estimated annual loss of $2 million for an industry valued at $323 million annually, the approximate value for the Delta region industry at the time, indicates a loss of less than 1 percent. While it is reasonable for the industry to want to minimize all losses, it is also important to recognize the scale of threat that cormorants posed to the industry. For individual ponds or producers, impacts can be far more significant, and this difference in scale has implications for how conflicts should be managed.

In the same special volume of *Waterbirds*, ADC published its review of techniques to prevent cormorant depredation at catfish farms, concluding that available options, at best, mitigated problems in the short term, but none of the measures alone or in combination were sufficient to resolve the conflict effectively. The USFWS also published a paper addressing its management responsibilities for cormorants relative to aquaculture, stating that it faced a "management dichotomy." On the one hand, it was the federal agency responsible for protecting fish-eating birds; on the other it was charged with promoting the development of private aquaculture. As a result, one of its primary goals was "to encourage aquaculture to develop in a manner that is compatible with responsible public resource stewardship. A component of that is to ensure responsible protection of migratory birds . . . while minimizing animal damage control problems at private aquaculture facilities." The agency acknowledged that "the take of limited numbers of birds will always have to be considered as a viable option in an effective, integrated program for minimizing the deleterious effects of fish-eating birds." It reconfirmed its commitment to work closely with ADC, state agencies, the aquaculture industry, and the conservation community in order to "realize an equitable solution to the problem of bird depredations at aquaculture facilities."[17]

The Aquaculture Depredation Order

In June 1997, the USFWS took a giant step forward in its attempt to realize a more "equitable solution" and proposed the first standing depredation order for cormorants. The first ever for a fish-eating bird, the order would do away with the need for producers to obtain a federal permit to kill cormorants at their ponds. It was expected to accomplish a variety of economic, social, and political objectives: reduced economic losses and damage-control expenses at aquaculture facilities, enhanced effectiveness of current nonlethal control programs, reduced paperwork and costs associated with the permit system, more efficient depredation-control operations as responsibility was shifted to individual aquaculturalists, cooperation between the federal agencies responsible for protecting and enhancing wildlife and for dealing with wildlife-damage issues, and cooperation with producers to address a problem that had "the potential to expand far beyond the wildlife management arena."[18]

The order proposed that fish farmers in thirty-two states, along with their employees or persons whom they designated as agents, be allowed to shoot *unlimited* numbers of cormorants "committing or about to commit depredations upon aquaculture stock." The USFWS anticipated that the order would bring about only a modest increase in numbers killed compared to numbers being killed under depredation permits. Nevertheless, the agency calculated a potential maximum kill of 92,400 cormorants yearly, a level of mortality, it argued, that would be more than offset by reproductive success and the recruitment of young birds into the population. The implication was that even if that many birds were killed, no real harm would be done if the viability of the population was not compromised.

The service received hundreds of comments supporting the proposal, along with many requests for additional flexibility in protecting aquaculture facilities. Farmers asked for authorization to use decoys, vocalizations, and other lures to bring cormorants into closer shooting range, and for the order to be expanded to include other species of fish-eating birds, such as egrets and herons, and to allow cormorant killing at roost sites. The only federal agency to submit comments was ADC. In August 1997, the program had rebranded itself yet again and would be known as USDA/APHIS–Wildlife Services, the more politically correct name it had donned in the 1960s in response to criticisms in the "Leopold Report." But its comments on the proposed depredation order indicated that while its name had changed, Leopold's recommendation for a policy embracing the "much broader responsibility for management of animal life in ways other than killing" had still not been adopted. Wildlife Services not only supported the proposed order, but also recommended widespread population control to reduce the size of the entire North American population. The agency suggested actions be undertaken at roost sites, on the breeding grounds, in western states, at sport fisheries, and at mariculture facilities, where marine animals are often cultivated for food directly in the ocean. The agency suggested that the unintentional or "incidental" taking of other birds be allowed, and that the certification requirement for producers—which specified that producers killing cormorants under the order had to have in place an established nonlethal harassment program approved by ADC—be deleted.[19]

Given that Wildlife Services had primary responsibility for developing and implementing depredation-control technologies, especially nonlethal strategies, these recommendations deserve close scrutiny. The cormorant-aquaculture conflict was defined by a limited seasonal and geographic scope, yet the agency proposed year-round control for the entire North American population, a continental reach for a localized problem. By including matters beyond the scope of the depredation order, such as control at sport fisheries on the northern breeding grounds, the agency made a case for the cormorant as a continent-wide problem, emphasizing a need to cull cormorants on a very broad scale. Wildlife Service's overreach on cormorant control, combined with its cavalier attitude toward the incidental taking of other birds and desire to eliminate certification requirements for producers, reveals a disturbing lack of restraint. Its recommendations recall the original criticisms of Rosalie Edge and the "Leopold Report," namely, that the animal-damage control program was indiscriminate and excessive, and had become an end in itself.

Only nineteen comments were opposed to or did not support the depredation order. Opponents expressed concerns that good data on the magnitude of economic impacts were lacking; that nonlethal measures had not been adequately implemented; that an effective strategy using nonlethal techniques and permits was already available; that the order would discourage aquaculturalists from investing in nonlethal, long-term solutions; that the geographic scope was unnecessarily broad; that the strategy did not address the spatially localized or seasonal nature of the problem; that the maximum number of birds that could be taken was way too high; and that the order set a dangerous precedent for other bird species. In addition, some commenters interpreted the proposed management actions as contrary to the purposes of the MBTA, an interpretation that resurfaced time and again from this point on.

On March 4, 1998, the USFWS published the Final Rule, establishing the first depredation order for double-crested cormorants.[20] The document includes the comments received and the agency's response, organized around twenty-five issues. Most requests and recommendations for changes to the order were denied, and lengthy explanations were provided in defense of many of the proposed actions. But a few significant concessions were made. One narrowed the geographic scope of the area covered under the order from thirty-two to thirteen states, twelve in the Southeast and one in the North, Minnesota. Another allowed aquaculturalists to use devices to lure cormorants into gun range. For the number of birds that could be taken, the agency noted that the maximum kill of 92,400 birds was a worst-case scenario. Instead, in the thirteen states where the order would be effective, 10,900 birds a year were being killed at the time the order was established, and the agency anticipated only a "moderate increase" from that figure. The order did not specify strict reporting requirements for producers, but did require producers to keep a log with dates and numbers of birds killed each month, to maintain each year's records for a total of three years, and to make the data available upon request to the USFWS.

The closest the service came to addressing how its proposed solution functioned equitably relative to its responsibility to protect cormorants under the MBTA was to state that "the MBTA provides considerable flexibility for dealing with situations where birds may come into conflict with human interests, such as the aquaculture-cormorant situation." It noted that there was a precedent for depredation orders that allow the killing of federally protected species without a federal permit. Conspicuously absent was any discussion of how the order would function equitably relative to cormorants themselves.

Finally, the USFWS reiterated a few of the more idealistic messages that it had presented in its 1995 paper: "The Service believes that the aquaculture industry shares responsibility for alleviating bird depredation problems and that the industry should aggressively promote: (1) The design of new facilities (and the retrofitting of old ones where economically feasible) that exclude or repel cormorants; and (2) the use of nonlethal deterrents. The Service also encourages Wildlife Services to continue an aggressive research effort to develop effective nonlethal means of alleviating bird depredation problems in aquaculture." But given the scope of the Aquaculture Depredation Order, such statements paid little more than lip service to fast-fading ideals. The order set a precedent in which the only factor limiting the number of birds killed would be the physical inability of a fish farmer to shoot more than a few birds at a time because of the birds quickly becoming wary, and in which any incentive to bird-proof facilities would fall by the wayside. With this order, the cormorant was on its way to becoming a feathered shadow of the coyote, an animal that remains, without a doubt, the poster child for predator control in America.

8

Conflicts on the Breeding Grounds

ON THE CORMORANT'S northern breeding grounds, most problems differ fundamentally from those on the southern wintering grounds. Conflicts in the North typically revolve around the cormorant's interactions with naturally occurring resources. Lakes, rivers, wetlands, estuaries, coastal bays, and other waters subject to an array of forces beyond human control are the usual places where conflicts occur. In those environments, the cormorant interacts with fish, plants, birds, and islands that belong to no one individual or industry, but to the realm of nature itself. In such a landscape, one might expect a greater tolerance for cormorants, since no personal ownership of resources is involved and interactions are driven by natural forces. Instead, human use of natural environments, particularly those prized for recreation and fishing, lays claim to most natural resources in a way that excludes large concentrations of cormorants. Indeed, for many, a sense of human ownership prevails even in the wildest places, and many natural lands, waters, and the life they support are held in state or federal ownership. They thus constitute public resources, a concept under which much wildlife is managed, and which would become especially important for cormorants.

At Michigan State University Bret Muter and colleagues examined media coverage of human-cormorant conflicts in the Great Lakes Basin, and their findings provide some important insights about the concept of public ownership of natural resources. Media coverage focused largely on the potential risks that cormorants pose to recreational activities such as fishing, and portrayed human-cormorant conflicts as public rather than personal issues. By contrast,

Leaping into a dive

conflicts on the wintering grounds emerge largely as problems affecting private property and business owners. But in both areas, the concept of human ownership is at the heart of the conflict. Interpreting the problem as a public one raises significant questions about where responsibility for the conflict and its resolution lies. In the Great Lakes, the media assigned responsibility for addressing public problems with cormorants to wildlife agencies and legislators.[1]

The Conflicts

Although issues north and south occur in distinctly different contexts, the key factor driving human-cormorant interactions in both locations is the same: cormorants consume fish in places where the fish are already spoken for. On the breeding grounds, a vast array of waters with fish suitable for cormorant consumption spans the northern half of the continent. As cormorant numbers increased, conflicts were eventually documented in most portions of the range. Often, no matter which species and what quantity of fish were consumed, cor-

morants were ultimately viewed as competing with human interests in one way
or another, either through direct predation on fish that people wanted to catch,
or through predation on the prey of the fish that people wanted to catch.

Some aquaculture production also occurs in the North, focused largely
on baitfish, food fish, and fish for stocking recreational waters. In Minnesota,
the only northern state included under the Aquaculture Depredation Order,
fish are typically stocked and grown in the state's abundant natural wetlands
and ponds, which are leased by fish producers from private landowners. Most
production occurs in the west-central portion of the state, an area overlapping
extensively with the breeding and migration ranges of cormorants, pelicans,
and other fish-eating birds. Like catfish ponds, the large and densely stocked
wetlands provide irresistible foraging opportunities. Here, producers consider
the use of exclusion devices and harassment practices even more impractical
because of the size and shape of the wetlands, their widespread distribution,
and ownership problems.[2]

A second important issue revolved around changes to island vegetation
due to cormorant roosting and nesting activities. A related concern was that
cormorants were outcompeting other birds for nesting habitat at particular
sites. These effects were perceived as direct impacts on humans because they
diminished the aesthetic quality of islands and caused ecological changes that
were considered negative. In several areas, such conflicts occurred simultaneously with those surrounding fish, and these combined impacts were later used
as further justification for cormorant control.

During the past thirty years, the most intense conflicts involving breeding birds occurred in the eastern and interior portion of the range. Although
conflicts occurred in the western half of the continent, too, birds did not become as abundant or problematic there. Thus, the remaining chapters will focus on conflicts and management east of the continental divide.

Coastal New England

In tracing the evolution of conflicts with cormorants in the late twentieth
century, Maine is an appropriate place to begin, both in terms of chronology
and because of the tension that developed there between the need to protect
cormorants on the one hand and the demand to manage them on the other.
Concerns about predation on Atlantic salmon in Maine led the US Fish and

Wildlife Service to issue the first depredation permit for cormorants in 1972, the same year that the birds became protected under the Migratory Bird Treaty Act. Some context for this apparent paradox can be found in the history of conservation efforts for the Atlantic salmon. Largely because of industrial and agricultural development and overfishing, many populations of this prized fish had been extirpated or greatly diminished in Maine as well as much of eastern North America by the mid-1800s. Hatcheries were established, and the Maine Atlantic Sea-Run Commission was created in 1947 to enhance and restore Atlantic salmon. Large numbers of salmon smolts were stocked in Maine rivers, and in the mid-1960s, evidence that double-crested cormorants were consuming stocked smolts was obtained at a few colonies in the eastern portion of the state. Although the effect of the predation was unknown, cormorants there and at other locations were shot, and an effort to eradicate them at one site through the destruction of eggs and young occurred annually until 1972.[3]

With the addition of cormorants to the MBTA, however, such activities required permission from the USFWS. At the time the cormorant became protected, the Maine Department of Marine Resources was spearheading the effort to reestablish the Atlantic salmon run to rivers in the eastern portion of the state. So that the agency could limit the potential impact of cormorant predation, it became the first recipient of a federal depredation permit for cormorants. Between 1972 and 1981, 2,799 cormorants were shot by wardens who were issued subpermits by the Maine Department of Marine Resources. But the cormorant population nonetheless increased, which led the state to alter its policy: in 1982, subpermits began to be issued directly to the public. As the decade progressed, additional fisheries data suggested that cormorant predation was not the main factor regulating the number of adult spawning salmon. But the possibility that cormorant predation could still be a factor affecting adult salmon return rates led to continued cormorant control. Between the late 1980s and early 1990s, fifteen to fifty subpermits were issued annually, and hundreds of cormorants were shot each year, mostly at rivers in eastern Maine. Agency time constraints resulted in the control program ending in 1993, although public demand for the program remained significant.[4]

Elsewhere on the coast, concerns developed that cormorants could displace other colonial waterbirds at particular locations prioritized for other species. Some local management occurred in Massachusetts and Rhode Island, but involved fairly small numbers of cormorants.[5]

Quebec

In Quebec, illegal control activities and complaints to governmental agencies regarding fish and vegetation impacts had been documented by the late 1970s at several locations. On islands in the St. Lawrence estuary in particular, cormorants were targeted, and biologists from the Canadian Wildlife Service noted that even though cormorants had gained protection through provincial legislation, it had not been possible to eliminate their persecution in that region. Fishermen convinced that cormorants were reducing fish stocks, along with island owners unhappy about the appearance of islands used by cormorants, provided an impetus for continued vigilante activity. Additionally, island owners consulted the Canadian Wildlife Service about their concerns over the cormorant's effect on vegetation. The agency responded by collaborating with landowners and Quebec's provincial authorities to once again initiate control activities at Île aux Pommes as early as 1978. For three years, all eggs and chicks were destroyed on the island, about a hundred adults were shot, and illegal control occurred too. Despite those activities, cormorants increased in the estuary, leading biologists to conclude: "Population control of cormorants can be achieved in the St. Lawrence estuary only by extensive, sustained and massive destruction of nests and their contents—an action that would justifiably cause indignation unless practiced with great discretion and using nonviolent means such as egg-spraying."[6]

Over the next decade, the estuary population continued to increase and the Ministry of Natural Resources received periodic complaints about a number of problems. Anglers and the sportfishing lobby complained about the cormorant invasion of salmon streams, and commercial fishermen reported problems with birds attracted to fish trapped in coastal weirs, such as sardines, smelt, and capelin. Additionally, local residents voiced strong concerns about impacts on forested islands. By the late 1980s, between 14,600 and 17,800 pairs of cormorants were nesting at twenty-five sites within the estuary, where there are about sixty islands. Twenty estuary islands were reported to be wooded, and several of those had been affected by cormorants since 1970. Some were described as having unique spruce and balsam fir forests that were "sometimes peculiarly unbalanced ecologically as a consequence of heavy browsing by herbivores in the absence of their natural predators." At four of the estuary's islands, birds nested entirely in trees, while at three they nested on the ground and in trees; the remaining sites were all colonized by ground-nesting birds. Predictions of expanding cormorant numbers led to increased concern about their impact on the islands, and a coalition of island owners lobbied the provincial government to adopt measures to prevent further damage from cormorants.[7]

In 1989, the Quebec provincial government initiated a five-year culling program. It noted that the desire to protect island plant communities, and not pressure from fishery sectors, was the reason for the cull. A population goal of 10,000 pairs was established, and birds at both tree- and ground-nesting colonies were targeted. During the first four years of the program, approximately 8,000 birds were shot and some 21,000 nests were oiled. By 1993, the last year of the program, the population had dropped to 9,561 pairs. Culling had reduced the population more rapidly than was anticipated, and the shooting effort was halted, although approximately 5,000 more nests were oiled. The scientists involved in running the program noted that "to this day, few populations of wild animals have been the subject of such extensive culls."[8]

Atlantic Canada

In Nova Scotia, conflicts in the 1970s led to an increasing number of requests to remove the cormorant's provincially protected status and to initiate population control. Interestingly, the complaints were identical to those received in the mid-1950s, with the exception of those regarding effects to vegetation. In the 1950s, complaints involved the destruction of trees on islands used for fishing purposes, navigation, and wood supplies, while those in the 1970s were described by the Nova Scotia Department of Natural Resources (NS DNR) as a "reduction in the perceived aesthetic quality of such areas." Fish-related conflicts in both periods included the destruction of Atlantic salmon, brook trout, commercial fish populations, and baitfish, along with removal of bait from lobster traps, damage to fishing nets, and the consumption of and damage to fish from fishing weirs. The original complaints had been investigated by H. F. Lewis, who found most were either inconsequential or were not supported by data. Lewis believed that a general dislike of cormorants was the basis for most complaints, but did acknowledge that cormorants could pose a local problem at some fishing sites. The department cited results of earlier studies showing that sport fish occur infrequently in the cormorant diet, and that the mostly marine component is variable and of little economic value.[9]

Recreational and commercial fishing interests did not accept those results and demanded a new study. In response, the department collected information on cormorant diet and foraging behavior during 1979–1989. Their results supported those of earlier studies, indicating that with the exception of impacts on fish in commercial weir and gill net operations, complaints could not be substantiated. They reiterated that cormorants were not destructive of commercial fish populations, and that most fish consumed by cormorants were of little value to either commercial or sport fishery interests. There was no data to support claims regarding impacts at lobster traps. Rather, cormorants were being accused simply because they were present in the area, and "the perception of a link was reinforced by the pronounced and rapid return of large numbers of cormorants foraging in the same areas in which fishermen set their traps."[10]

The other Maritime Provinces where cormorants increased also received complaints, especially related to fisheries. In 1996, birds were illegally shot and harassed at a large colony on Prince Edward Island, and although the identity of the perpetrators was not discovered, conflicts with fisheries there suggest that fishermen were probably responsible.[11]

The Prairie Provinces: Lake Winnipegosis

In the Prairie Provinces, increases in cormorant numbers led to significant fishery-related conflicts in many locations, but nowhere has the level of conflict and illegal destruction of cormorants been greater than on Manitoba's Lake Winnipegosis. During the period of initial population recovery in the 1970s, cormorants experienced a brief respite from illegal control. But the belief that cormorants were responsible for commercial fishery declines remained alive, despite the fact that the walleye fishery had collapsed during the 1960s, when cormorant numbers were greatly diminished. By 1979, the increase in cormorants in the province was again leading to pressure on provincial authorities from commercial fishermen, joined by trout farmers, to help eliminate the "crowduck menace."[12]

By 1986, declines in the walleye harvest on Lake Winnipegosis were so significant that a three-year commercial fishing closure was initiated. The following year, the Manitoba Department of Natural Resources (MB DNR) undertook a cormorant diet and population study to examine the role of cormorant predation in the walleye decline. Results indicated that while cormorants had increased greatly on the lake, the commercially valuable walleye and sauger comprised a mere 0.3 percent of their diet, while three forage species, suckers, perch and tullibee, made up 94 percent of what cormorants were eat-

ing. The results supported conclusions of earlier studies, reiterating that cormorants seldom consume commercially valuable species. The study also revealed that human persecution was the most significant source of mortality for
cormorants in 1987. At fifteen of the thirty-seven active colonies, birds were
found shot, nestlings were crushed under rocks, and eggs were destroyed. The
department provided the diet study results to provincial fisheries scientists, and
articles were written for the Manitoba commercial fishermen's newsletter.

Despite these efforts, persecution continued. A paper summarizing the
1987 study reported additional observations of colony attacks in the following
year: "Techniques employed by fishermen in 1988 included the brief heating
of eggs with a flamethrower device. . . . More common techniques involved
the destruction of eggs and young typically through clubbing. Shotguns were
used to kill adults and to destroy young in tree nests."[13] In 1989, disturbance
was documented at eleven colonies, with thousands of chicks and birds destroyed. On one island, the eggs and young of all gulls, in addition to those
of the cormorants, were thrown into the water. In 1994, a massive slaughter
at multiple colonies was discovered by the USFWS biologist Ken Stromborg
and provincial authorities. Stromborg was in the area to collect a small number
of birds for pollution-related studies. In addition to the destruction of some
4,000–6,000 cormorants, approximately 2,000 white pelicans were destroyed,
the latter species perceived by fishermen to share a role in commercial fish
declines. The story was featured in the Canadian journal *Birds of the Wild* and
included graphic pictures documenting the event. Stromborg, commenting on
the scale of the killing, stated that he was "unaware of any other instances of
such concentrated destruction of migratory birds in North America."[14] Approximately two years later, the scientist, environmentalist, and broadcaster
David Suzuki hosted a powerful documentary called *Cormorants and Pelicans:
Prairie Scapegoats* for the CBC's *The Nature of Things*, which publicized the vi-
olence and controversy surrounding these two species on Lake Winnipegosis.

Despite the attention given to this persecution, a complete survey of colonial waterbirds on Lake Winnipegosis in 1999 documented that large-scale
persecution of cormorants remained a prominent activity on the lake. Survey results indicated that most accessible cormorant colonies along the main
travel lanes were systematically disturbed. The wildlife biologist Bill Koonz
with the Manitoba Department of Natural Resources, who was in charge of the
survey effort, reported that the disturbance displaced cormorants from tradi-
tional nesting sites, causing them to colonize multiple new and forested sites.

Additionally, he relayed the following powerful story. While attempting to count nests on Dog Island Reef, Koonz and his staff encountered three men; two were armed and shooting birds on nests where chicks were hatching. As Koonz approached, they continued to shoot, and although their actions were illegal, they were reluctant to leave the island. The men stated that the cormorants were ruining the island, which had previously been forested. Koonz reported some 500–1,000 dead adult cormorants lying on the island's shoreline in addition to the ones shot on their nests. Furthermore, many of the young birds that had managed to escape the shooters because of Koonz's interruption nevertheless did not escape death; with their parents kept from the nests by the shooters, the birds rapidly broiled in the sun.[15]

The Great Lakes Region

Considering the tremendous cormorant population growth that occurred in the Great Lakes between the late 1970s and early 1990s, problems developed surprisingly slowly there. The first conflict was documented at Lake Superior's Apostle Islands, where commercially valuable whitefish began declining in the early 1970s. Cormorants recolonized the area in the late 1970s, and by 1980 the area's single colony had grown to fifty-three pairs. Commercial net fishermen began to complain about cormorant depredations at their nets. By 1982, the colony had approximately quadrupled in size, and five fishermen reported a loss of 30–40 percent of their whitefish catch to cormorant predation and fishing activities. But a diet study indicated that cormorants there mostly consumed forage fishes and that lake whitefish occurred in only 1 percent of the cormorant diet sample obtained. Observations suggested that pound nets were attractive to cormorants as perch sites, and that the presence of cormorants increased the rate at which fish were being gilled, a process in which fish become trapped in the net when the mesh slips behind their gills. The problem resolved itself in the mid-1980s when the combination of a depressed whitefish market and the presence of cormorants led commercial fishermen to abandon pound nets and switch to trap nets.[16]

Elsewhere in the region, no evidence of renewed conflict was recorded until 1984 when an act of persecution was observed on New York's soon-to-be-infamous Little Galloo Island in the eastern basin of Lake Ontario. An important smallmouth bass and recreational fishery contributed substantially to the regional economy and way of life, especially in Henderson Harbor, New

York, a rural community on the lake's eastern shore. A visit to the island in July found that in a colony of 736 pairs of cormorants, large rocks had been placed on top of birds and eggs, and hundreds of chicks lay dead in their nests.[17]

The same year, the first nesting records were obtained on New York's Oneida Lake, some twenty miles southeast of Lake Ontario and long recognized for its important fish resources. Intense fishing pressure, combined with numerous environmental changes, led to dramatic changes in the lake's fish composition, and populations of many predator species declined. Walleye, however, benefited substantially, and since the 1940s this species, along with yellow perch, has dominated the fish community. The sport fishery contributes significantly to the region's economy, and although no conflicts were initially documented when cormorants arrived, Oneida Lake would soon become a locus for such problems.

Three years later, two more vigilante acts were observed, this time on Lake Michigan's Little Gull and Gravelly Islands in Delta County, Michigan. The discovery was made by James Ludwig while collecting cormorant diet data on Lakes Huron, Michigan, and Superior to assess the birds' impact on commercial and sport fisheries. In May 1987, visits to the two islands revealed that all eggs from hundreds of cormorant nests had been destroyed, while nests of Caspian terns and gulls had been left alone. At the time, Michigan's cormorant population consisted of about 2,200 pairs. Later, results of the diet study reported that the cormorant diet varied seasonally and geographically, and was dominated by small forage fishes, especially alewife. Valued species, such as yellow perch and smelt, made up a somewhat substantial portion of the diet only in the Georgian Bay area of Lake Huron. The study concluded that cormorants were not important competitors of commercial fishermen, but were generalist predators preying upon the most abundant and easy-to-catch fish available, typically small forage fishes.[18]

By the late 1980s, conflicts were becoming more frequent and widespread in the Great Lakes region. In the eastern basin of Lake Ontario, the smallmouth bass fishery had begun to decline. Anglers and business owners complained that the quality of the fishery had deteriorated, and residents from Henderson Harbor formed the group Concerned Citizens for Cormorant Control. Meanwhile, the cormorant population on Little Galloo Island continued to increase. By 1992, 5,443 pairs were estimated, and the New York State Department of Environmental Conservation initiated a five-year study to assess fish losses to cormorants in the eastern basin, working in conjunction with

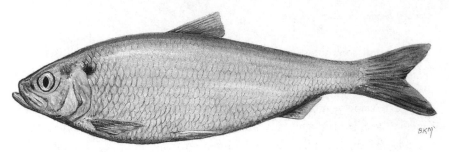

Alewife, a nonnative fish in the Great Lakes that was very abundant when
cormorants began recovering there

the US Geological Survey. In June of the same year, another vigilante-style
killing of cormorants was documented on Little Galloo: more than a dozen
adult cormorants were found dead along the island's eastern shore, apparently
shot from a boat at close range.[19]

Concurrently in Ohio, the first successful nesting of the cormorant in
over a century was documented on West Sister Island in western Lake Erie.
This forested island, which is designated as both a National Wildlife Refuge
and a Federal Wilderness Area, provides habitat for several colonial waterbird
species. Concern developed almost immediately that cormorants would affect
vegetation and displace the other waterbirds. Similar concerns developed in
the New York and Vermont waters of Lake Champlain, and the New York and
Toronto waters of Lake Ontario.[20]

Evidence obtained in 1992 showed that persecution was also occurring
in Canadian waters. A trip to West Island in Georgian Bay, Lake Huron, documented what became an annual shooting party to destroy cormorants at that
nesting location. In 1993, a wholesale slaughter of chicks and the destruction
of nests on Herbert Island in the North Channel, Lake Huron, was observed.
Additionally, fifty adult cormorants were the victims of an illegal shooting on
Lake Ontario's Pigeon Island in the eastern basin.[21]

The first authorized cormorant control actions on the Great Lakes since
the bird obtained protected status were undertaken in New York in the mid-
1990s. On Little Galloo Island, cormorants continued to increase, and the
state's Department of Environmental Conservation became concerned that
they would expand to nearby islands valued for their vegetation and the habitats that they provided for other colonial waterbird species. To prevent those

islands from being colonized, the state obtained a depredation permit from the USFWS and destroyed small numbers of eggs and nests on nearby islands. Likewise, the Vermont Department of Fish and Wildlife and USDA Wildlife Services began destroying eggs and nests on Lake Champlain's Mud Island to prevent cormorants from colonizing new islands.[22]

About the same time, Michigan's Les Cheneaux Islands in Lake Huron began to emerge as another important locus for cormorant conflicts. For decades, the area had supported a significant yellow perch fishery and a recreation-based economy, but in the late 1970s, anglers and local citizens became concerned that the fishery was declining. Anglers first identified commercial fishing by Native Americans, increased predation by introduced species of fish such as brown trout and chinook salmon, and heavy angling pressure during the spawning period as possible reasons for their diminished catches. Cormorants did not begin to reestablish themselves in the area until 1980, and then remained in relatively small numbers throughout the decade; thus, they were not initially implicated in the declines. Size limits on yellow perch were initiated in 1987, but did not lead to recovery. Concurrently, cormorant numbers increased from about 700 pairs in 1989 to 4,000 pairs in 1995. A study was then initiated to evaluate the effect of cormorants on the area's yellow perch fishery.[23]

Meanwhile, more vigilante activity occurred. Two domestic pigs were released on an island in Green Bay, Wisconsin, ostensibly in the hopes of eliminating cormorants. But cormorants, which nested in trees at the site, mostly ignored the pigs, while the real victims were ground-nesting herring gulls and red-breasted mergansers. The pigs altered the island's vegetation, and the herring gulls temporarily abandoned the site. Over the winter the pigs died and the following spring their skeletons were found lying side by side.[24]

In the midst of these developments, the USFWS produced a fact sheet in May 1995 titled "Cormorants and Their Impacts on Fish." The first sentence stated: "In evaluating impacts on fish populations, it is important to distinguish between perception and reality." The fact sheet went on to point out that "studies conducted worldwide have repeatedly shown that the amount of fish consumed by cormorants in natural situations totals much less than 10 percent of the quantity caught by commercial or sport anglers." The service noted that it was reviewing the impact of cormorants on commercial and sport fish populations, and was considering a variety of options to address real and perceived problems. The fact sheet concluded by stating that "based on a review of the

best scientific evidence, it does not appear that a strategy of reducing cormorant populations to benefit sport fish populations is biologically warranted,"
although the agency would maintain an "open mind on the subject."

Concerns in the region continued to mount, and in 1997, the USFWS and
the Canadian Wildlife Service convened a symposium to discuss management
issues related to cormorants in the Midwest. Many of the papers presented were
published in 1999 in an APHIS technical bulletin. Some highlights of the findings: The Les Cheneaux Islands study reported that mortality sources other
than cormorant predation were responsible for the removal of most legal-sized
yellow perch. The study of the eastern basin of Lake Ontario reported that
the cormorant diet over a five-year period consisted of only about 1 percent
game fish. For western Lake Erie, diet work indicated that cormorants were
not detrimental to sport or commercial fishing. At Ohio's West Sister Island,
monitoring indicated that cormorant numbers had increased and that vegetation damage had occurred, while the number of black-crowned night-herons
and great blue herons had declined. But declines in night-herons were linked
mostly to succession changes in vegetation, and causes for great blue heron
declines were unknown.[25]

The USFWS presented a paper, and relative to what soon thereafter
came to pass, the paper is particularly noteworthy. The agency presented the
results of an extensive review of cormorant diet studies undertaken in freshwater habitats between 1923 and 1994, work that confirmed what it had suggested in its 1995 fact sheet: cormorants generally have minimal impact on
sport fish populations. The service restated that a strategy of large-scale population reduction to benefit sport fish was not biologically warranted, and that
it currently did not issue cormorant depredation permits to benefit sport fish
populations in public waters. On the matter of vegetation impacts, the agency
recommended that depredation permits be issued on public lands only if there
was convincing evidence that endangered or threatened plant species, or rare
or declining plant communities, were being harmed by cormorants. To protect
other birds, the USFWS would issue permits only if convincing evidence indicated that cormorants were harming endangered or threatened species or a
regionally significant population.[26]

To conclude this section, the year 1998 and the state of New York provide a
most appropriate time and place to end. Events here ultimately brought the
"cormorant problem" on the northern breeding grounds into sharp focus.

They marked a turning point in cormorant management and once again high-lighted the tension between the need to protect the birds and the demand to manage them.

On July 29, Little Galloo Island was yet again the site of vigilante activity, but this time the enormity of the crime could not be overlooked. During the 1990s, Little Galloo had become the site of the largest-known cormorant colony on the continent, reaching a high of 8,410 pairs in 1996. The colony then began to decline to an estimated 5,839 pairs by 1998. Despite the decline, threats against the birds began building early in the year. In April, threats materialized, and 8 cormorants were illegally killed. Then, in June, another 100 were illegally destroyed. Finally, in July, a massive shooting event killed some 850 birds, wounded more than 100, and left countless chicks to bake in the sun or starve in their nests. The magnitude and cruelty of the slaughter, noted as one of the largest mass killings ever of a federally protected species, were so great that the event momentarily captured the nation's attention. The large-scale destruction was reported in the *New York Times*, which published a feature article covering the grim spectacle on August 1.[27]

Less than ten days later, the *Times* ran a second in-depth feature on the massacre and reported on the attitudes of people living in Henderson Harbor, where cormorants were viewed as a threat to fishing, jobs, the economy, and the way of life. The article reported that the locals likened the slaughter to the Boston Tea Party and hailed the unknown perpetrators as heroes.[28]

In the fall, the first official actions to protect sport fish from cormorants were launched on Oneida Lake. Preliminary data from cormorant-diet and fishery-monitoring studies suggested that cormorants were having an effect on walleye and yellow perch, leading Wildlife Services and the state Department of Environmental Conservation to initiate an intensive hazing program to disperse migrant cormorants from the lake. Initial actions did not involve killing birds, but they mark the beginning of what would become a significant management program. Later studies published on cormorant diet and impact on Oneida Lake would prove to be pivotal for cormorant management.

Finally, as the year came to a close, an event occurred that would seal the cormorant's fate for years to come. In mid-December, the Department of Environmental Conservation completed its series of cormorant diet and impact studies in the eastern basin of Lake Ontario and compiled the results in an un-published special report. Findings indicated that cormorants were responsible for a measurable decline in sport fish, and that predation on smallmouth bass in particular was "excessive."[29] While the degree to which the studies were able to reliably support these conclusions was not beyond question, a matter discussed in detail in chapter 12, this assessment nevertheless came to be considered "proof" that cormorants indeed diminish fish resources valued by humans.

Thus, the year ended with a high-profile dilemma for the agency charged with "protecting, restoring and managing" migratory birds for "the benefit of the American people." In the case of the double-crested cormorant, the US Fish and Wildlife Service contemplated a responsibility that had become a seemingly impossible contradiction in terms.

9

The Second Standing Depredation
Order for Cormorants

It was a bright cold day in April, and the clocks were striking
thirteen.

—George Orwell, *Nineteen Eighty-Four* (1949)

AS THE MILLENNIUM drew to a close, a portrait of a troublesome, destructive bird emerged. Three events in particular—the attack at Little Galloo, the establishment of the Aquaculture Depredation Order, and the results of New York State's diet studies—set into motion a force that was of such magnitude and direction that it acquired a life of its own. Like a disease to which few had any immunity, the belief that cormorants were a pestilence upon humans and nature alike spread rapidly across the bird's range, infecting nearly all those whom it touched. Despite numerous studies from multiple areas indicating that cormorants were not negatively affecting fish resources, evidence from just one location alleging that cormorants were causing sport fish declines confirmed what many had believed all along. Almost immediately, a new rash

Preening

of cormorant studies was launched, and there was seemingly no limit to the crimes the cormorant could commit.

Over the next decade, hundreds of articles on cormorants were printed or aired. So began a media frenzy that introduced the American public to a truly nefarious character. Time and again, the cormorant was described as an "ugly, black" bird with a "voracious" appetite and a "foul fetid" smell, a creature that "bedeviled" its environment. The magnitude of damage associated with its habits and behavior, along with its generally evil nature, were comparable only to such end-of-the-world events as the apocalypse or a nuclear blast. In the Great Lakes basin, the species received such significant coverage that Bret Muter and colleagues undertook the research cited in the previous chapter, a review of some 140 newspaper articles written between 1978 and 2007. Published in 2009, the opening clause of the study's title, "From Victim to Perpetrator," sums up how perceptions of the cormorant were transformed as it recovered in the Great Lakes.[1]

A less well documented but no less potent force was also contributing to the cormorant's deteriorating image. Social networking, particularly through fishing forums, communicated mostly erroneous information and promulgated hatred for the bird. On websites such as FishingBuddy.com, FishingMinnesota .com, and many others, anglers vented about how ugly, useless, and disgusting they found cormorants to be. Contributors posted recommendations for how to do away with the problem, and occasionally confessed to "accidentally" destroying cormorants while out shooting vermin. Additional research by Muter and colleagues indicates such forums are important sources of information sharing. Muter interviewed a network of people involved in cormorant management, including agency professionals, anglers, bird watchers, and tourist industry personnel, regarding their preferred sources and channels of information about cormorants. Results indicated that social networks provided the greatest source of cormorant-related information for the majority of those interviewed and were powerful sources of influence.[2]

As the bird's evil reputation spread across the land, wildlife agencies began receiving numerous telephone calls from the public expressing concern about the bird's potential to destroy. Fish, plants, birds, water quality, air quality, private property, airplanes (via collisions), human health, and entire ecosystems were all perceived to be threatened. Sometimes people called simply because they had seen a cormorant on a lake—an observation they believed some authority should be aware of and probably do something about. Concurrently, myriad initiatives, resolutions, legislative bills, position statements, and less formal prescriptions were developed in the United States. Suggested fixes ran the gamut from potentially viable strategies to radical approaches that required dismantling the bird's legally protected status. The numerous proposals in the latter category typically reflected the interests of particular stakeholders and, though often preposterous, amplified the sense of urgency for the USFWS. Together, the calls for action and the wide range of suggested treatments exerted a tremendous influence on the development of cormorant policy in the United States.

Soon, many government documents emerged, making declarations about and proposals for cormorants in a nearly indecipherable language of acronyms, agency jargon, legalese, and bureaucratic divisions. Nevertheless, studying them is fruitful because it reveals the politics behind cormorant management and the forces that led to the second depredation order for these birds.

Pressures Far and Wide

First to voice a congressional opinion and to suggest one of many controversial approaches to cormorant management was the Lone Star State. On February 25, 1999, Texas state representative Ron Clark introduced House Concurrent Resolution (HCR) 23 on the double-crested cormorant. Unlike a legislative bill, a resolution does not become a law, but is often used to formally express a legislative opinion. HCR 23 identified the cormorant as a "water turkey," a "sky rat," and a "fire ant with wings," and analysis of the resolution contributed the description "fish coyote." The resolution described how the cormorant had "flocked to Texas in growing numbers to winter in the warmer climate," and stated that "its feeding habits and increasing numbers are decimating sport fish populations in many areas . . . [and] fish raised by commercial aquaculture facilities." It asked the 76th Texas Legislature to pass a resolution to request that the US Fish and Wildlife Service "include the economic impact of the cormorant on sport fishing in its ongoing evaluation of the double-crested cormorant and . . . consider the removal of the double-crested cormorant from the protection of the Migratory Bird Treaty Act if the national economic losses warrant severe control methods to keep the cormorant population at a manageable level within a given region." Three months later, HCR 23 was signed by then Texas governor George W. Bush.[3]

As that resolution was being considered, Vermont and New York initiated a more direct approach for resolving their problems with cormorants. On March 19, the USFWS received a request from the Vermont Department of Fish and Wildlife to implement a large-scale cormorant reduction program on Lake Champlain. The management goal was to enhance and protect diversity on the small, state-owned Young Island and elsewhere in the Vermont waters of the lake. Cormorants colonized Young Island in the early 1980s, and by the time Vermont submitted its request, some 2,600 pairs were nesting there. That development allegedly led to the island's abandonment by black-crowned night-herons, cattle egrets, and snowy egrets after trees that those species were nesting in became occupied by cormorants. Specifically, Vermont requested permission to prevent all successful reproduction within Vermont waters of Lake Champlain and to prevent cormorants from establishing colonies on any other state-owned islands. Proposed activities included oiling all eggs, killing nestlings, and administering avian contraceptives to adults on Young Island, and destroying nests on all other islands.[4]

Three days later, a similar request was received from the New York State Department of Environmental Conservation to reduce cormorants in the eastern basin of Lake Ontario on Little Galloo Island. The management goals included improving the eastern basin smallmouth bass fishery, enhancing fishing opportunities, reducing cormorant impact on vegetation and increasing nesting opportunities for other bird species. Regarding vegetation, when New York submitted its request, other species present on Little Galloo were all ground nesters, and included one of the largest ring-billed gull colonies on the continent, estimated at 54,000 pairs, a sizable and growing population of Caspian terns, and smaller numbers of herring and great black-backed gulls. These species were not believed to be negatively affected by cormorants, and in fact the gulls may have benefited from cormorants dropping fish and supplying food in the form of eggs and young. Thus, benefits, if any, would accrue to other bird species from managing cormorants on Little Galloo Island only by preventing cormorants from expanding to other islands, a policy that was already in place through earlier permits.

To reach its goals, New York proposed a five-year plan that established the first of many population objectives for cormorants. The target for Little Galloo was 1,500 pairs and a 90 percent reduction in nesting success, a goal based on the maximum numbers present on the island before declines in smallmouth bass were observed. The state requested permission to oil as many as 7,500 cormorant eggs and to shoot as many as 300 cormorants as an experimental precursor to the period between 2000 and 2002, when the state anticipated shooting an unspecified number of birds to reach its population objective of 1,500 pairs by 2003. Because birds at Little Galloo represented about 60 percent of New York's population in the late 1990s, this objective entailed not only a massive reduction of the island's population, but also of the state's.[5]

Approximately two and a half weeks after the USFWS received these requests, ten men from sport-fishing communities near Little Galloo pleaded guilty to charges related to the 1998 shooting on the island. The story ran in the *New York Times* on April 9. Of the five who pleaded guilty to the most serious charges, one was the founder and leader of Concerned Citizens for Cormorant Control. Each was sentenced to six months of house arrest, fined $2,500, and agreed to contribute $5,000 to the National Fish and Wildlife Foundation. The other five, three of whom were sons of the leader of Concerned Citizens for Cormorant Control, pleaded guilty to lesser charges. The three brothers admitted to previously shooting cormorants on Little Galloo in separate incidents, as did a fourth, unrelated man whose job during the most recent shooting had been to hide the weapons used by the shooters. The five men pleading guilty to the more significant charges had participated in the earlier shootings as well. For their crimes, the three brothers and the man who hid the weapons agreed to serve three months of house arrest and to pay smaller fines. Finally, an additional man who admitted to being present but not to doing any shooting pleaded guilty to being an accessory to the crime and was fined and placed on probation.[6]

Several of the guilty men were professional fishing guides or marina owners, and throughout their plea negotiations, they continued to assert the belief that the need to save their livelihoods had justified their actions. And although New York's request to manage cormorants had not yet been approved by the USFWS, the men were aware of it. They reportedly felt "vindicated" by the state's plan to cull cormorants on Little Galloo, seen primarily as an effort to protect sport fish. The defendants' lawyer further interpreted the state's plans as proof that killing the birds had been the appropriate action for the men to take, telling the media, "They've done nothing other than what the State of New York is going to do itself."[7]

The local fishing community and business owners responded with strong shows of support for the criminals. Anti-cormorant paraphernalia, including T-shirts, baseball caps, pendants, and bumper stickers, were sold to raise money for their case and to help pay their fines. Additionally, businesses hosted fund-raising dinners and donated merchandise for raffles. The men received letters praising their actions—one reported receiving hundreds.[8]

Within a month of the guilty pleas in the case, the USFWS issued final environmental assessments for cormorant management in Vermont and New York.[9] For each state, the depredation permits were amended to allow large-scale egg oiling, but the requests to kill adults and nestlings and to use contraceptives were denied. Relative to New York's request, the service identified the current federal policy for cormorants relative to fish, stating that it did not issue permits for controlling cormorants feeding in natural environments unless they were feeding on fish that were endangered, threatened, or part of a restoration effort. Thus, management was allowed only to benefit other birds and vegetation; but the USFWS noted that this management focus could help reduce predation on bass. The population objective based on the timing of smallmouth bass declines went unchallenged, and in fact, cormorant management specifically to benefit fisheries became a well-acknowledged aspect of the state's program from then on. Losing no time, New York and Vermont began oiling eggs immediately. That summer, 99 percent of the 5,681 nests present on Little Galloo were oiled, and hatching success for oiled eggs was less than 1 percent. Similarly, thousands of nests were oiled on Vermont's Young Island, and nearly all reproduction was eliminated.[10]

In the developing universe of cormorant control, these large-scale management efforts emerged as highly significant. They represented a return to the cormorant management philosophies of the mid-twentieth century and set a precedent for management in the twenty-first. In August 1999, a lawsuit was brought against the USFWS by the Atlantic States Legal Foundation, charging that the decision to issue a permit for control on Little Galloo was "arbitrary and capricious" and violated both the MBTA and the National Environmental Policy Act. However, the court found in favor of the USFWS. Meanwhile, the management actions at Little Galloo did not make problems with cormorants go away. Despite the program's success in preventing thousands of chicks from hatching, the residents of Henderson Harbor derived little satisfaction from the egg-oiling effort. The community continued to market its anti-cormorant trinkets and donated proceeds of the sales to Concerned Citizens for Cormorant Control. The group's leader, empowered by community support and bolstered by his faith that killing cormorants was the solution to the fishing community's problems, remained outspoken on the issue.

As the year progressed, Concerned Citizens for Cormorant Control caught the eye of New York congressman John McHugh, who grew up in Watertown, New York, just sixteen miles west of Henderson Harbor. Sympathetic to the community's problems, McHugh introduced HR 3118 in the fall of 1999, a bill authorizing states to establish a hunting season for double-crested cormorants. He publicly endorsed Concerned Citizens for Cormorant Control, stating that the group "could be of invaluable assistance to our efforts. Contact your fellow sportsmen, fishing, and hunting groups throughout the nation and ask them to join us in this battle." The congressman was highly critical of the USFWS, claiming that the agency's

> failure to adequately address the cormorant crisis in the eastern Lake Ontario basin has been one of the most frustrating experiences I have encountered since coming to Congress in 1993. . . . Despite the years of studies and stalling, they have completely failed to meet their responsibility to design and implement an effective cormorant management strategy. In the face of FWS continued refusal to help us address the grave threat posed by these birds I have introduced cormorant hunting legislation. In addition, I am writing separate legislation amending the Migratory Bird Treaty Act to enable the Secretary of the Interior to issue regulations authorizing the Fish and Wildlife Service to take action when cormorants pose a public nuisance and/or deplete natural resources.[11]

Several states to the west, in Minnesota, Congressman Collin Peterson was in the midst of his own cormorant problem. By the time McHugh introduced HR 3118, thousands of cormorants were once again nesting in the remote Lake of the Woods. Despite findings by the Minnesota Department of Natural Resources that fish populations were stable or increasing, fishermen believed that cormorants were negatively affecting walleye and sauger populations. Congressman Peterson, who represents the northwestern portion of the state, where fishing and hunting interests are prevalent, became the cosponsor of HR 3118. In the spring of 2000, Peterson commented to the press: "The cormorant population in northern Minnesota has literally exploded, devastating the game fish population and causing environmental havoc. . . . If the Fish and Wildlife Service does not act swiftly to manage and control this destructive bird, then Congress will. . . . Personally, if the cormorants were wiped out, I would not think that was a bad thing."[12]

Meanwhile, in the southern portion of the cormorant's range, a new problem was developing. Around the time when New York and Vermont launched their egg-oiling programs, Arkansas initiated a cormorant control program at its only known breeding colony, on Millwood Lake. The location was probably colonized by cormorants in 1989, representing the first documented nesting in Arkansas since 1951. On May 19, 1999, the colony was visited by personnel from the Arkansas state office of Wildlife Services, and more than one hundred cormorant nests were documented. During the visit, Wildlife Services shot eleven birds, purportedly to obtain data for DNA testing, food habits, and other studies. On June 30, Wildlife Services returned to the lake and shot 106 of the 137 birds present, again to collect data for DNA and food studies. But Arkansas Wildlife Services' state director, Thurman Booth, who soon emerged as a prominent and controversial figure in the cormorant battles, was quoted in the *Catfish Journal* as saying that his agency hoped to "keep this population down and maybe eliminate it." Neither the catfish industry nor sports fishermen, he explained, wanted cormorants to become established year-round in Arkansas, although the presence of nesting birds represented a recovery of the bird's former status as a breeder in the state.[13]

Similarly, in the Delta region of Mississippi, two small breeding colonies discovered in 1998 constituted the only known nesting by cormorants in the state since 1952. The observations were documented in a paper jointly published by Wildlife Services and USFWS biologists, and the authors reported that unknown persons had removed some or all of the nests at both locations. Although the cormorant's return as a breeding bird to Mississippi was a return to the bird's former status in the state, the same concerns were voiced there as in Arkansas: "If breeding populations continue to increase throughout the region," the authors noted, "the aquaculture industry may have to contend with cormorant depredation not only as a seasonal occurrence but as a year-round issue."[14]

A Widening Schism, Some Additional Players, and an Orwellian Approach to Management

These were just some of the cormorant-related pressures placed on the USFWS in 1999. The cormorant problem had gained a seemingly unstoppable momentum, and public dissatisfaction with the birds and criticism of the federal agency managing them could not be ignored. For many months, the USFWS had been contemplating a national management plan for the species but had not officially announced that idea. In early 1999, the agency had enlisted researchers at the University of Minnesota to prepare a comprehensive status assessment for the double-crested cormorant in North America as a precursor to developing such a plan.[15] As the year came to a close, the agency published a "Notice of Intent" in the *Federal Register*, stating that it would prepare an environmental impact statement (EIS) and a national management plan to address cormorant impacts. The public was invited to submit comments and attend public scoping meetings around the country. In its approach to cormorant management, the service noted that it might consider a range of options, from "No Action" to "Large-scale Population Control on Breeding Grounds, Wintering Grounds, and Migration Areas in the United States."

The preparation of an EIS indicated that significant changes in policy and major actions were on the horizon for cormorants. An EIS requires an agency to describe the proposed action, provide a range of alternatives to it, analyze the environmental impacts associated with each potential action, and consider public comment. It is a complicated document that requires a large investment of time and other resources by a federal agency. In the EIS prepared for cormorants, completed four years later, the USFWS identified specific National Environmental Policy Act guidelines that triggered its decision to prepare the document. These included undertaking a "major federal action" and "precedent-setting actions with wide-reaching or long-term implications."[16]

Although not indicated in the initial announcement, the USFWS had made the somewhat controversial decision to prepare the EIS and management plan in cooperation with Wildlife Services. Since population reduction and lethal control largely define the latter agency's past and current approach to resolving conflicts with wild animals, the alliance was another indicator that something major was on the horizon for cormorants. The stated reasons for the partnership were the increasingly important role Wildlife Services was taking on in cormorant management and research, and the need to foster cooperation between the two agencies, particularly in coordinating future cormorant management strategies.

While the EIS was being prepared, other developments arose that would influence the politics around cormorant management. The first was a legislative change that was not made specifically to address cormorant conflicts, but nevertheless had an important bearing on the agencies managing the species. In July 2000, the US Court of Appeals for the District of Columbia Circuit ruled that the prohibitions of the MBTA applied to all federal agencies, over-turning 1997 circuit court rulings that had exempted agencies such as Wildlife Services from permit requirements when resolving damages related to non-game migratory birds. After the 2000 ruling, all federal agencies were once again required to obtain a federal permit to kill nongame migratory birds, and the USFWS immediately notified state and federal agencies that it interpreted its MBTA-associated regulations as applying to each of them.

The ruling led to further controversy over cormorant management and widened the schism some had perceived between the USFWS and Wildlife Services since 1986, when Animal Damage Control was transferred back to the USDA. Arkansas, in particular, viewed the ruling as a move by the USFWS to strip Wildlife Services' authority to manage birds. The issue of cormorant control at the Millwood Lake colony brought the significance of the ruling to the fore. In 2000, a few months before the ruling was made, Arkansas Wild-life Services had again shot out the Millwood Lake cormorant colony, but the new permit restrictions resulting from the ruling meant that from then on the agency would need USFWS approval before taking any such action. The ensu-ing conflict between Arkansas Wildlife Services and the USFWS was covered in detail by the Delta Farm Press, which repeatedly quoted Arkansas Wildlife Services director Thurman Booth. Booth reported that after the 2000 ruling, the USFWS told his agency that "if [it] shot them for another year [it would] be violating the Migratory Bird Treaty Act. Their contention was that there was no data to indicate those nesting birds were causing problems." While Booth acknowledged that this was essentially true, he indicated that it was only a mat-ter of time before the birds would cause problems. From his perspective, the issue of cormorant control at Millwood Lake highlighted the larger problem involving the authority to manage birds and served as a prime example of how the FWS "meddled" in Wildlife Services' business.[17]

As this controversy was brewing, important developments emerged from the Wildlife Services' National Wildlife Research Center in Starkville. First, Jim Glahn and his colleagues published a revised estimate of replacement costs for fingerlings consumed by cormorants in the Mississippi Delta, indicat-ing that costs had approached $5 million by the late 1990s, more than doubling since 1990. Shortly thereafter, Glahn gave a highly influential presentation (later published) at a USDA National Wildlife Research symposium. Based on new studies (described in chapter 14), Glahn and his colleagues estimated that losses of catfish, if calculated at harvest, could be five times greater than the costs to replace fingerlings, and approach a whopping $25 million, almost 9 percent of all catfish sales.[18]

One month later, the National Wildlife Research Center issued a comprehensive document titled "A Science-Based Initiative to Manage Double-crested Cormorant Damage to Southern Aquaculture." The initiative advocated ideas previously offered by Wildlife Services in its comments on the proposed Aquaculture Depredation Order. Although the focus was on limiting impacts to southern aquaculture, the initiative made a case for addressing the cormorant problem at a flyway scale, which encompasses the path traveled by birds between the breeding and wintering grounds. Management at that scale would target a population using a specific flyway (for example, the Mississippi or Atlantic flyway) at all places and during all phases of the annual cycle (breeding, wintering, and staging grounds). The document was presented as part of an integrated strategy to reduce cormorant impacts on southern aquaculture, but also addressed the multitude of other problems allegedly caused elsewhere by cormorants. The authors stated that "without human intervention, breeding populations in the Great Lakes will likely continue to increase, resulting in more habitat destruction, competition with other colonial waterbirds, competition with sport fishermen, and depredations on southern aquaculture farms." Although Wildlife Services lacked the authority to implement its recommendations, the initiative provided a well-defined strategy that presented large-scale population reduction as the logical conclusion.[19]

Meanwhile, the debate over agency authority for managing migratory birds continued to pick up steam. In May 2001, Arkansas senator Blanche Lincoln introduced Senate Bill S909, which had two objectives: exempt Wildlife Services from the requirements of the National Environmental Policy Act if it was undertaking any migratory bird management activity, and provide authority for any agent or employee of Wildlife Services to issue permits to stakeholders and cooperators to take migratory birds. The bill would not only have provided duplicate authority to manage birds, but by exempting Wildlife Services from the requirements of NEPA, would also have gone beyond any flexibility possessed by the USFWS. Essentially, such exemption would allow Wildlife Services to kill cormorants without considering the environmental impact of such actions. The bill was cosponsored by Arkansas senator Tim Hutchinson. Four months later, an essentially identical bill, HR 2879, was introduced by

Arkansas representative Mike Ross. The House bill had four cosponsors, from Arkansas and Mississippi. Both bills were referred to committees, and an executive comment was requested from the USDA for the House bill.

A month after the first of these bills was introduced, yet another important player emerged on the scene, the Great Lakes Fishery Commission. Established between Canada and the United States in 1955, the commission's two major responsibilities are to develop research programs and recommendations that result in the maximum sustained productivity of Great Lakes fish stocks, and to formulate and implement programs for the eradication or reduction of the parasitic sea lamprey in the Great Lakes. For each lake, committees help implement the commission's strategic plan, address specific issues, and help shape management policy for Great Lakes fish. In June 2001, the Lake Ontario committee passed a resolution identifying the cormorant as a problem for the entire Great Lakes Basin. Citing the New York studies as reason for concern about fish impacts, the committee recommended that the commission "take an active role in advocating and supporting policies and measures for the management of the Double Crested Cormorant, where their numbers are shown to impact the fishery and the ecosystem."[20]

The following month, Wildlife Services drafted and formally adopted a position statement on cormorants. The agency advocated population reduction nationally, regionally, and locally to reduce impacts on aquaculture, natural resources, property, and human health and safety; its strategy was to use "all efficacious damage management methods at nesting, roosting, wintering and all other applicable sites where cormorants are found." In comments made to the Delta Farm Press, the ever-outspoken Thurman Booth noted his part in drafting the statement, along with counterparts from Louisiana, Mississippi, Alabama, and Florida.[21]

On December 3, 2001, the USFWS draft EIS for cormorant management became available to the public. Twelve public scoping meetings had been held at locations across the country, and some 1,450 written comments had been obtained. The document identified six potential strategies to address conflicts with cormorants, ranging from no action to developing a regional population-reduction program or hunting program. It was noted that biologists from Wildlife Services' National Wildlife Research Center advocated the population-reduction strategy, but that the USFWS, which retained ultimate regulatory authority, proposed a strategy that involved expansion of the Aquaculture Depredation Order to allow lethal control at night roosts, and the development of a second standing depredation order, to be called the Public Resource Depredation Order. The order would authorize state fish and wildlife agencies, federally recognized tribes, and federal land management agencies to implement cormorant management where they deemed cormorants injurious to public resources, including fish, wildlife, and plants.

To understand the significance of the strategy chosen by the USFWS, consider that the Public Resource Depredation Order, by definition, transformed all natural and common elements the cormorant interacts with during its day-to-day life—fish, other birds, and plants—into highly valued public resources that could be guarded from cormorants. Since cormorants eat fish, manipulate vegetation, and compete with other species in their daily activities, it is hard to imagine a scenario in which some degree of injury to "public resources" does not occur. With this new strategy, the USFWS not only reversed its policy on cormorants feeding in natural areas, but also, in a truly Orwellian fashion, redefined fish, trees, and other birds—essentially "nature"—in a way that could make them off limits to cormorants in just about any situation. Moreover, by granting authority for cormorant control to other federal, state, and tribal agencies, it allowed the USFWS to pass the responsibility for management decisions. In these ways, the proposed depredation order set important new precedents for the management of a fish-eating, tree-nesting bird and opened a whole new world of possibilities for cormorant management.

With the federal government's preferred strategy identified, the writing was on the wall for cormorants. But before a national management plan could be put into place, several more rounds of public comment, additional scoping meetings, the publication of three more documents, and nearly two more years were required. Of the many opinions expressed during that time, two are included here because they represented the attitudes of wildlife professionals who were peers of the FWS, and continued to track and provide insight into significant issues.

In February 2002, the Wildlife Damage Management Working Group of the Wildlife Society posted a draft position statement on the USFWS's proposed management plan. Since its formation in 1937, this society of professional wildlife biologists has acquired a large international membership and multiple working groups, and has developed close to fifty position statements on wildlife issues. The Wildlife Damage Management Working Group recommended that the USFWS work with other agencies and use a combination of strategies to reduce regional or continental cormorant populations. It suggested using "the most efficacious lethal control methods, in conjunction with existing prevention and harassment techniques, to . . . reduce the overpopulation of cormorants as quickly as possible." In comments on the bills proposed by the Arkansas legislators, the working group viewed their development as a failure by agencies to adequately manage cormorants: "The U.S. Fish and Wildlife Service should take more aggressive actions to manage regional and local cormorant overpopulation and resulting damage caused by this species. We believe that the lack of adequate management progress has . . . led to a loss of credibility of wildlife management agencies and prompted legislation proposed in the Congress (Bills S. 909 and H.R. 2879) to address the Double-crested Cormorant overpopulation concerns. Both bills, if passed, would force wildlife agencies, in which the bills' authors and sponsors apparently have lost confidence, to aggressively address the problem."[22]

The month after this position statement was posted, Thurman Booth made several inflammatory comments to the Delta Farm Press. He identified important philosophical differences between Wildlife Services and the USFWS regarding goals, each agency's willingness to kill animals, and an alleged institutional jealousy on the part of the USFWS. Regarding how the cormorant issue had been handled to date, he stated: "The Fish and Wildlife Service has been a travesty and is unethical. They've shirked their responsibility." As for the proposed management plan, he stated: "I don't think Fish and Wildlife Service's goals are honorable, so I don't think they're doing anything sound. There's no doubt of that. Having known the people who are suggesting this, having understood their philosophy and having been an employee of that agency for over 20 years, I know this is the case. . . . What they're trying to do with this plan is continue to cater to PETA and the Humane Society and the rest." Conversely, he interpreted the measures proposed by the Arkansas legislators as an effort to provide his agency with the authority needed to manage cormorants, stating that Wildlife Services was receptive to the proposed legislation and that it would help the agency to resolve problems. Alluding to a provision in the National Environmental Policy Act that allows a "categorical exclusion," he argued that a precedent for exemption from the act's requirements had already been set.[23]

In the world of cormorant politics, a continual supply of such comments has established Thurman Booth as a controversial figure. In 2010, a visit to his office revealed a cormorant nest on display at the entrance, and a man proudly proclaiming it to be the only one of its kind in Arkansas, thanks to the perseverance of him and his agency.[24] Although Booth is given to outlandish statements, a little history indicates that his views provide a window into the philosophic differences and ongoing power struggle between the two principal federal agencies attempting to manage these birds. For twenty years before the program's transfer to the USDA, Booth worked in animal damage control as an employee of the USFWS. In comments to the Delta Farm Press, he described a "philosophical shift" that took place within the USFWS during his tenure there, a time when the program was experiencing strong criticism from committees convened by the US Department of the Interior. He asserted that a no-kill philosophy slowly infected the agency and that "many individuals within the Fish and Wildlife Service aren't hunters and philosophically disagree with the concept of killing anything." But once animal damage control was transferred back to USDA, according to Booth: "We found those in management atop us were oriented toward food and fiber production. They understand that in order to have a steak for supper, something has to be killed. Fish and Wildlife Service is currently kowtowing to people ignorant to that fact."[25]

Booth's main contention was that the USFWS prevented Wildlife Services from accomplishing its tasks. "We've had the full responsibility to deal with cormorants, but not the authority," he noted, a sentiment harking back to the late 1980s, when the newly restructured Animal Damage Control program first encountered "obstacles" to resolving conflicts with migratory birds. From Booth's comments, it is clear that the "obstacle" in the way of cormorant management had become the USFWS itself. That sentiment is echoed in the legislation proposed by the Arkansas legislators, comments of the New York and Minnesota legislators, the position statement of the Wildlife Society's Wildlife Damage Management Working Group, and in many less formally documented opinions. Additionally, giving Wildlife Services more authority to manage cormorants was one of the suggested management options identified in public comments on the draft EIS, mainly advocated by people in the Southeast.

In attributing the philosophical differences between the two agencies to the differing orientations of the USDA and the Interior Department, Booth went straight to the heart of the problem in agency conflicts over cormorants. The mission of the latter is to "protect and manage the nation's natural resources and cultural heritage," while that of the USDA is to "provide leadership on food, agriculture, natural resources, rural development, nutrition, and related issues." Theoretically, those missions should not be mutually exclusive, but fundamentally different objectives, agendas, and means are often apparent. In this context, the partnership between the USFWS and Wildlife Services for cormorant management is quite significant and poses an ongoing question: is it possible to arrive at a synthesis of these agencies' objectives for cormorants, or do the values and goals of one ultimately trump those of the other?

The Final Environmental Impact Statement and the Final Rule

The next official benchmark in the plan's development was publication in the *Federal Register* (March 2003) of a proposed rule for cormorant management. Despite calls for more aggressive management through regional population reduction and the continued criticism of the USFWS, the proposed rule continued to advocate the preferred options identified in the draft EIS. The expanded Aquaculture Depredation Order and the newly developed Public Resource Depredation Order moved another step closer to becoming the official strategies for addressing cormorant conflicts. One notable change from the draft EIS was that under the second order, the only federal agency granted authority to implement cormorant management would be Wildlife Services. A sixty-day comment period was set up to allow public opinion again to be expressed, and during that time, an unfathomable 9,700 comments were received.

While the proposed rule was being reviewed, the final environmental impact statement (FEIS) was completed, and a notice of its availability was published in August 2003. The document identified the Public Resource Depredation Order and the expanded Aquaculture Depredation Order as the proposed alternative plans for cormorant management. The nearly 1,600 comments submitted on the draft EIS indicated only weak support (15 percent of commenters) for the proposed alternative recommended by the USFWS. The FEIS indicated that although Wildlife Services continued to support "a population reduction strategy based on scientifically determined population levels and social carrying capacity," the proposed alternative was preferable because of uncertainty about the need for large-scale population reduction and its ability to reduce resource conflicts, and because population reduction would require significantly more resources than currently existed for research, monitoring, and coordination.[26]

The last step in the process was the publication of the final rule on October 8, 2003, establishing the regulations to implement the expanded Aquaculture Depredation Order and the Public Resource Depredation Order.[27]

Although thousands of comments had been received since these strategies were initially proposed, few truly substantial modifications occurred between their conception and implementation. In the final rule, the USFWS organized and condensed the nearly 10,000 comments it received on the proposed rule into fifty-seven items that it considered significant enough to respond to. The agency addressed the majority of comments by restating its belief that it was acting responsibly relative to concerns being raised, the MBTA, and its mission. The agency challenged several of the concerns as being either the responsibility of other agencies or simply incorrect, and on others argued in defense of its position. A handful of concerns resulted in modifications to the proposed policy, but these did not change the nature of the options chosen. A thirty-day waiting period followed to allow the public and agencies to become familiar with the new regulations. Finally, on November 7, 2003, the second standing depredation order for cormorants and the new policies it established went into effect. Eleven days later, Wildlife Services issued its official "Record of Decision," which stated that as a cooperating agency, it was adopting the FEIS, but reiterated that its preferred alternative included regional population reduction as well.

10

A Half Million and Counting

Implementation of Management Policies in the United States

FOR CORMORANTS, the newly devised standing depredation orders created an environment not seen since the nineteenth century. Though cloaked in legalese, rules, and regulations, they established such loose requirements for killing and destruction that once again this black interloper became the target of intense persecution. Across much of the eastern United States, feeding areas, roosting sites, and breeding colonies were transformed into killing fields. Nesting islands, where birds demonstrate the strongest attachment to place and concentrate in great numbers, provide especially poignant reflections of times gone by. For just as before, when islands were the frequent destination of market hunters, plume collectors, eggers, and fishermen, so again they have become the target of those seeking birds and eggs.

The exact number of cormorants legally destroyed in the United States since the first depredation order was established in 1998 is not known, but reasonable estimates indicate that by the end of 2011 the number had exceeded some half million birds (table 10.1). In addition, an untold number of nests and eggs were destroyed.

Wounded cormorant at a cull

The Aquaculture Depredation Order

Of the three existing management policies—depredation permits and the two standing depredation orders—most cormorants have been killed under the Aquaculture Depredation Order. Between 1998 and 2010, the last year for which data were obtained, hundreds of thousands had been killed under that order alone (table 10.1). Although the order targets cormorants mainly on the wintering grounds in very specific circumstances—birds at or in the vicinity of aquaculture facilities—these locations are where many cormorants spend part or all of the winter season. In addition, few legal restrictions limit the numbers of birds that can be killed. The only condition producers must meet in order to take birds under the order is to have a nonlethal harassment program in place as certified by Wildlife Services. Other limitations are that they kill birds only with firearms and only during daylight hours. Considering that nothing more formal is required, it comes as little surprise that large numbers of cormorants were destroyed during the order's first thirteen years.

The 1998 order did not specify strict reporting requirements for producers, so numbers killed in the initial four years (1998–2001) were based on estimates obtained through a survey that the USFWS conducted with fish producers. According to the data, nearly 130,000 cormorants were killed at aquaculture ponds during the first four years the order was in effect. The fi-

Table 10.1 Cormorants killed under the Aquaculture Depredation Order (AQDO), the Public Resource Depredation Order (PRDO), and some depredation permits, 1998–2011

Year	AQDO	Permits (P) or AQDO & permits (AP)	PRDO	Total
1998	31,734	12,484 (P)	0	44,218
1999	39,698	12,385 (P)	0	52,083
2000	36,190	10,493 (P)	0	46,683
2001	22,000	21,669 (AP)	0	43,669
2002	19,736		0	19,736
2003	22,292		0	22,292
2004		25,203 (AP)	2,395	27,598
2005		22,005 (AP)	11,221	33,226
2006		30,947 (AP)	21,428	52,375
2007		26,052 (AP)	19,960	46,012
2008	16,560		18,782	35,342
2009	15,002		25,562	40,564
2010	14,501		18,363	32,864
2011	NA		28,389	28,389
Totals	217,713	161,238	146,100	525,051

Sources: 1998–2001: USFWS 2003a (FEIS); 2002–2003: USFWS unpublished data, provided by T. Doyle; 2004–2007: Permits and AQDO from USFWS 2009, PRDO from USFWS unpublished data, provided by T. Doyle; 2008–2011: USFWS unpublished data, provided by T. Doyle.

nal environmental impact statement (FEIS) included data on the number of birds killed under depredation permits during this period, and these data are included here because in some later years the number killed under depredation permits was not separated from the number killed under the order. Depredation permits resulted in more than 57,000 cormorants being taken, bringing the total killed between 1998 and 2001 to more than 186,000 birds. For the years 2002 and 2003, no survey was conducted, but a review of other data collected by the USFWS produced an estimate of 42,000 birds killed under the order during these two years.

The expanded order of 2003 provided a greater range of circumstances in which birds could be killed. In addition to daytime shooting at ponds, birds could be shot at roost sites day or night, anytime from October through April, either by Wildlife Services personnel or by those designated as their "agents." The broader scope resulted in more rigorous reporting requirements, and both producers and Wildlife Services are now required to submit data on the numbers of birds they kill at farms and night roosts. For the years 2004 through 2007, combined data from killings under the order and under depredation permits indicate more than 104,000 cormorants were killed. For the years 2008 through 2010, an estimated 46,000 cormorants were killed under the order alone. This brings the total number of birds killed under the order between 1998 and 2010, along with some birds killed under depredation permits, to approximately 379,000. Although the exact proportion attributable to the order can't be determined from these data, it probably accounted for upward of 80 percent, or about 300,000 birds.

How close these estimates come to the actual number of birds killed at and in the vicinity of aquaculture ponds over the years is anyone's guess. For example, the average annual number of birds killed based on producers' reports is lower than the average based on the survey data, but the survey data are thought to be more representative. Furthermore, it is highly likely that some number of birds were killed but not reported. Referring to the frustration experienced by catfish farmers, Thurman Booth told the Delta Farm Press, "Let's be honest, there are probably more birds killed illegally than under permits." Thus, the reported estimates are probably best considered the minimum number killed.[1]

Birds taken under the order have been killed in all thirteen states where the order is in effect, but data were not obtained on the numbers of birds killed in each location. For the period between 2004 and 2007, most birds were killed in Mississippi, Alabama, and Arkansas. Given the distribution of aquaculture facilities and documented sites of conflict, it can reasonably be assumed that these states accounted for much of the mortality during the entire period. Minnesota, the only northern state included under the order, killed approximately 23,350 cormorants between 1998 and 2010—only about 8 percent of the total killed under the order, but still a substantial number of cormorants.[2]

The Public Resource Depredation Order

Compared to the Aquaculture Depredation Order, the Public Resource Depredation Order has a far wider reach, encompassing not only the thirteen states making up most of the cormorant's migration and wintering ranges, but also eleven more states in the Midwest and East where significant cormorant breeding populations occur. Additionally, because this order is directed toward birds allegedly affecting public resources, birds are targeted mostly on public rather than privately owned lands. This distinction is significant because many of the public lands where birds are targeted have been set aside precisely because of their importance to the conservation and protection of waterbirds and other wildlife, an issue discussed in more detail in chapter 16. Because public lands provide much of the breeding habitat for cormorants, control at most of the sites supporting large numbers can be implemented under the second depredation order if agencies so desire.

This order also encompasses a broader range of conflicts and control techniques than its forerunner, and birds judged to be negatively affecting a public resource can be shot, killed by cervical dislocation, or asphyxiated with carbon dioxide. In addition, the eggs and nests of the birds can be oiled or destroyed. Moreover, the order allows agencies to reduce or even eliminate local breeding populations if they comply with specific requirements and evaluate the effect of management on cormorants and the resources potentially affected. Besides the personnel of state fish and wildlife agencies, federally recognized tribes, and Wildlife Services, people designated as agents of these agencies have the authority to carry out such actions.[3]

The requirements to authorize lethal control activities are easy to meet, and no federal permit is necessary. Before any control activities can be initiated in a given year, all agencies must submit a one-time written notice to the appropriate Regional Migratory Bird Permit Office. If an agency plans to remove 10 percent or less of a colony, it need only inform the USFWS of its intentions and provide some justification for management. If more than 10 percent of a colony is targeted, a written notice thirty days in advance of the activity is required, and must include details regarding location, proposed activity, resources being affected, the total percentage of cormorants to be killed, and which other species of birds are present. Reporting requirements, which are intended to help the USFWS maintain federal oversight of cormorant management, include

keeping records of activities and providing an annual report with a summary of
birds killed and eggs oiled; a description of the impacts or anticipated impacts
that cormorants are causing to public resources; a description of the evidence
supporting the conclusion that cormorants are causing or will cause these impacts; a discussion of other factors affecting the resource; and a discussion of
how control efforts are expected to, or actually did, alleviate the impacts.

Importantly, the order is intentionally devoid of guidelines for what constitutes an impact, how to distinguish real from perceived impacts, or how to
document the significance of an impact. Under its terms and conditions, the
order requires "a description of the evidence supporting the conclusion that
double-crested cormorants are causing or will cause these impacts," but there
are no specifications for the evidence to be based on scientific inquiry or rigorous data. In response to a public comment on the order that guidelines for
what constitutes an impact should be issued, the USFWS stated that it "wanted
to maximize the flexibility of other agencies in determining what constitutes a
public resource depredation. We understand that there are concerns about all
of the 'what ifs' that could conceivably take place in the absence of guidelines.
We have made the purpose of the depredation orders clear, and we trust that
our agency partners will not abuse their authority." Even more significant is
that agencies' evidence in support of alleged impacts and of how control will
(or did) alleviate them needs to be described in the report only *after* the control
activity has taken place.[4]

Management under the order began in 2004 and has targeted mostly
breeding birds. Data on numbers killed through 2011 indicate that the order
was responsible for the destruction of 146,100 cormorants during this eight-
year period. Combining these numbers with data from the Aquaculture Depredation Order, along with some limited data from depredation permits, indicates that the Public Resource Depredation Order accounted for about
28 percent of the birds killed between 1998 and 2011 (based on the assumption
that numbers killed under the Aquaculture Depredation Order in 2011 were
equal to the three-year average for 2008–2010; see table 10.1). While this is
not a large proportion, it represents killings over a shorter time period; over
the last several years, killings under the second order have surpassed those
under the first. Furthermore, killings under the PRDO are coupled with massive losses in reproductive efforts. For instance, during the first four years the
second depredation order was in effect, more than 111,000 nests were reported
oiled or destroyed.[5] Since that time, additional egg-oiling programs have been

implemented at many locations. By 2011, it is likely that at least as many nests were destroyed again, and probably far more. Given a typical clutch size of four eggs, and assuming at least 250,000 nests were oiled between 2004 and 2011, a minimum of a million cormorant eggs have probably been destroyed. As with the first order, the permissiveness of the second all but guaranteed that large numbers of cormorants and their eggs would be destroyed in places where high levels of conflict and negative perceptions about the birds prevail.

Control under the second depredation order has in many ways been far more controversial than control under the first. The players include a range of state, federal, and tribal agencies allegedly managing for the public good; under the first order, most killing is done by private individuals, overseen by a single federal agency managing birds to protect business interests. Under the second order, most birds are destroyed at or near their breeding colonies, rather than killed when they show up to feed at fish farms. As a result, the Fund for Animals, the Humane Society of the United States, Defenders of Wildlife, and the Animal Rights Foundation of Florida brought a lawsuit against both the USFWS and Wildlife Services to challenge the new rules. But the Court eventually found in favor of the governmental agencies.[6]

The more complex environment for which the second order was designed immediately gave rise to two significant events that foreshadowed what was to come. The most significant was the process that unfolded for cormorant management on the breeding grounds, particularly in the Great Lakes Basin. In May 2004, the first of several environmental assessments for large-scale management that implemented the order was authorized for Michigan. Similar to an environmental impact statement though less detailed, an environmental assessment must identify and analyze proposed actions and alternatives. Interagency review may lead to a conclusion that no further analysis is required and the agency(s) may proceed with the proposed action, or conversely, that a full-scale EIS should be prepared to examine the potential actions.[7]

As the only federal agency authorized to control cormorants under the second order, Wildlife Services was heavily involved in the process that developed. In addition to having technical expertise, its direct contact with stakeholders experiencing problems with cormorants, and its mission to provide leadership in managing wildlife conflicts, gave the agency a key advisory and coordination role. Consequently, the agency took the lead in preparing the environmental assessment for Michigan, while the USFWS participated as a cooperator. The proposed action was to initiate a cormorant management program that implemented the second order and included local population reduction. On May 21, 2004, the proposed action was authorized to proceed.

From all vantage points, the management program that emerged in Michigan was remarkable. It immediately established Michigan as the state taking the most aggressive action across the breeding grounds. Previously, New York had taken the lead on cormorant management, and while it continued to implement and expand an intensive management program, the bulk of its efforts consisted of egg oiling rather than killing adult birds. Michigan would go on to destroy five times as many cormorants as New York and annually remove far more than several states combined (table 10.2). While management under the second depredation order got off to a relatively slow start, with only about 2,400 cormorants killed in just five states during the first year, Michigan claimed 59 percent of the birds destroyed, and oiled eggs in more than 3,000 nests. The program initially targeted breeding cormorants in the Les Cheneaux Islands area, along with spring migrants at a few locations to disperse birds from sites valued for fishing. In addition to Wildlife Services personnel, volunteer citizens were designated as agents to assist with cormorant harassment activities.[8]

While this bureaucratic approach to cormorant management was unfolding, a much more casual but nevertheless significant one was developing in Texas. On August 30, 2004, the Texas Parks and Wildlife Department issued a press release announcing, "Local areas in Texas besieged by the double-crested cormorant, a federally-protected bird more commonly referred to as the water turkey, can get depredation relief under a new Texas Parks and Wildlife Department control permit program." The permit, officially known as the "Nuisance Double-crested Cormorant Control Permit," could be had for $12 by any private landowner in Texas with a valid hunting license. Permit holders and their designated agents could then kill cormorants on privately owned land identified in the permit, via firearms, cervical dislocation, and carbon dioxide asphyxiation. The only reporting requirements are that permit holders provide an annual summary of the number of cormorants killed and maintain records for two years.[9]

Since the double-crested cormorant can be taken only through federally issued depredation permits or in compliance with depredation orders, the Texas nuisance permit presents an interesting set of circumstances. Through a liberal interpretation of the second depredation order, which was established

Table 10.2 Reported number of cormorants killed under the Public Resource Depredation Order, by state, 2004–2011

State	Year								Total
	2004	2005	2006	2007	2008	2009	2010	2011	
Alabama		1,143	1,523	700	122	348	98	45	3,979
Arkansas	254	134	145	30	41	27			631
Florida									0
Georgia	30								30
Illinois									0
Indiana					1	1			2
Iowa				50	50	5			105
Kansas									0
Kentucky									0
Louisiana									0
Michigan	1,421	2,429	5,627	7,767	8,223	9,768	7,119	8,093	50,447
Minnesota		2,793	3,103	2,461	2,601	3,084	2,222	1,848	18,112
Mississippi			633	697	426	458	303	218	2,735
Missouri									0
New York	482	1,665	1,924	1,669	1,418	1,808	601	1,215	10,782

(continued)

Table 10.2 (*continued*)

State	Year								Total
	2004	2005	2006	2007	2008	2009	2010	2011	
North Carolina									0
Ohio			5,873	3,725	2,664	2,357	2,206	3,727	20,552
Oklahoma					8				8
South Carolina									0
Tennessee									0
Texas		2,599	2,272	2,636	2,500	5,286	4,309	6,046	25,648
Vermont	208	458	328	222	503	1,001	631	4,000	7,351
West Virginia									0
Wisconsin				3	225	1,419	874	3,197	5,718
Total	2,395	11,221	21,428	19,960	18,782	25,562	18,363	28,389	146,100

Source: USFWS unpublished data, provided by T. Doyle.

to protect public resources, a state agency was able to issue a nuisance permit for a federally protected bird on private lands. This elastic interpretation of the second order was made possible by the USFWS's willingness to go beyond the order's intended purpose and to introduce a new area in which benefits to public resources could potentially occur. In a letter dated August 9, 2004, the USFWS authorized Texas to pursue the permit and provided the following detail on how such action was plausible under the second order:

> Although ambiguously treated in the depredation order, control of Double-crested cormorants by private landowners to allevi- ate depredation losses to their private resources is not intended. Rather private individuals . . . are expected to apply for Federal depredation permits to control Double-crested Cormorants that are impacting their private resources. However, double-crested cormorants undoubtedly depredate without bias on both private and public resources on both private and public ownerships. Ac- cordingly, it may be presumed that control efforts to alleviate depredation impacts to one ownership category benefit the other, especially if those ownerships are in close proximity. There- fore, if TPWD [Texas Parks and Wildlife Department] deter- mines that public resource interests will be served by controlling double-crested cormorants on private lands/waters, they or their agents may implement control activities as long as landowner per- mission is received.[10]

With this authorization, the nuisance permit became the process by which Texas Parks and Wildlife designated agents for cormorant control.

In 2005, more agencies and tribes got on board. Wildlife Services pre- pared a second environmental assessment for cormorant control in the Great Lakes Basin, this one for Minnesota. Cooperators included the USFWS, the Leech Lake Band of Ojibwe, and the Minnesota Department of Natural Re- sources. Most of the management was planned for Leech Lake, part of the Leech Lake Indian Reservation and home to the largest cormorant colony in the state. Management goals were to protect the lake's walleye and yellow perch fisheries, and its small colony of about 200 pairs of common terns. A cormo- rant population objective of 500 pairs was identified; achieving it would mean an 80 percent reduction, or the removal of 4,000 birds. During the first year of the program, about 70 percent of the birds targeted were killed (table 10.2). In Wisconsin, no significant management occurred, but the state legislature passed an act requiring the Wisconsin Department of Natural Resources to administer a cormorant management plan, making cormorant management in Wisconsin a state law. And Texas began reporting what became consistently large numbers of cormorants killed under the nuisance permit (table 10.2).[11]

The following year, an environmental assessment was prepared for another Great Lakes basin state, Ohio, with the USFWS, the Ohio Department of Natural Resources, and the West Sister Island National Wildlife Refuge as cooperators. This assessment was particularly significant because it marked the first time that a wildlife refuge joined forces in large-scale cormorant reduction. Initiated on Lake Erie islands and inland lakes to protect biodiversity from potential cormorant impacts, the program established population objectives encompassing colony reductions ranging from 48 to 100 percent, and statewide reductions of 49–57 percent. These goals required the removal of thousands of breeding birds, and in the program's first year nearly 6,000 cormorants were destroyed (table 10.2). In Michigan, an amendment to the 2004 environmental assessment was prepared, and the National Park Service came on as an additional cooperator. This amendment authorized an expansion of management activities in order to protect fish in several more areas, and to protect vegetation at South Manitou Island, part of Sleeping Bear Dunes National Lakeshore. It indicated that Michigan's cormorant population, then estimated at about 30,000 pairs, could be reduced by as much as 50 percent over time. The number of cormorants killed annually would be allowed to increase to 9,700 birds a year, more than twice the number predicted for individual states by the FEIS.[12]

By 2007, control efforts in the US Great Lakes had become widespread, and the majority of cormorant colonies in the region had experienced some form of management at least once. Additionally, several tribal authorities in Michigan had begun controlling cormorants in northern Lake Michigan and the St. Mary's River as part of the larger initiative undertaken by Wildlife Services and other agencies. In addition to some killing, the tribes oiled eggs and destroyed substantial numbers of nests.[13]

In 2009 and 2010, additional environmental assessments were prepared for Wisconsin and Vermont. Cooperators in Wisconsin included the USFWS's Migratory Bird Management Office, the Wisconsin Department of Natural

Resources, and the Horicon, Gravel Island, and Green Bay National Wildlife Refuges. Population objectives requiring large reductions were developed to reduce potential impacts to fish, vegetation, and other birds, and the initial strategy was to use primarily egg oiling and to achieve reductions gradually. For the Vermont assessment, which was a supplement to one from 2004 for overall bird damage in the state, cooperators included the Migratory Bird Management Office and the Vermont Department of Fish and Wildlife. The supplement identified a need to increase the number of cormorants killed annually, suggesting that Wildlife Services could take up to 4,140 cormorants a year to alleviate cormorant damage and threats.[14]

In 2011, a need for even more aggressive management was identified in Michigan, and the environmental assessment for that state was revised yet again. In addition to earlier cooperators, the Michigan Department of Natural Resources and staff from multiple tribal authorities served as consultants in the document's preparation. The maximum number of cormorants that could be killed annually was increased to 15,500, and a minimum of only 5,000 breeding pairs was required to be maintained in the state. Population objectives were established for multiple areas, requiring the reduction of many thousands of birds. Wisconsin likewise shifted its approach and began killing substantial numbers of cormorants.[15]

To summarize: between 2004 and 2011, a handful of states distinguished themselves as major players in the world of cormorant management (table 10.2). In Michigan alone, more than 50,000 birds were killed, and tens of thousands of nests were oiled or destroyed, nearly all in the hopes of protecting fish. About half as many cormorants were killed in Texas, allegedly for the same purpose. In Ohio, more than 20,000 cormorants were shot to protect birds and trees at a handful of sites; in Minnesota, more than 18,000 cormorants were killed, mostly to protect a few hundred common terns and a sport fishery at one of the state's ten thousand lakes. New York distinguished itself by destroying tens of thousands of nests and developing the most aggressive egg oiling and nest destruction program to date; and Vermont and Wisconsin emerged as up-and-coming participants in the process to reduce cormorants. Most of this work was carried out by Wildlife Services personnel, but staff from natural resource agencies, tribal authorities, and wildlife refuges were involved also, along with an undisclosed number of citizens designated as agents in Michigan and Texas.[16]

A Note on Costs

Cormorant control is expensive, not only in bird life lost but in dollars spent as well. An idea of the expense can be gleaned from the data in table 10.3. In general, expenses are justified on the assumption that cormorant control is cost-

Table 10.3 Some operational (O) and monitoring (M) costs associated with cormorant control in the United States

Location	Time period and costs	Funding sources	Source
Les Cheneaux Islands, MI	2004–2008: $155,000 (O)	Not provided	Fielder 2010b
Les Cheneaux Islands, MI	1995: $100,000 (M)	Not provided	Diana 2010
Northern Lake Huron	2010: $1,000,000 (M) estimated to annually monitor this area	Not available	Diana 2010
Michigan (statewide)	2011: $125,000 / yr (O)	Federal and state	USDA 2011-MI
Leech Lake, MN	2005–2012: $850,000 (O, M)	Leech Lake Band of Ojibwe, Bureau of Indian Affairs, USDA, MN DNR	Steve Mortenson, Leech Lake Band of Ojibwe
Oneida Lake, NY	1998–2005: $3,500,000 (O)	Federal	Shwiff 2009
Great Lakes (multistate)	2011–2012: $175,000 / yr (O) (Total $350,000)	Great Lakes Restoration Initiative	USDA/APHIS– Wildlife Services[a]

a. See http://greatlakesrestoration.us/projects/aphis.html.

effective relative to the economic gains supposedly realized by its undertaking. But little work has been done to support this justification. In fact, a formal analysis of costs and benefits has been published for just one control program, that at Oneida Lake, New York. Undertaken by the National Wildlife Research Center, the analysis reported economic losses of hundreds of millions of dollars and thousands of jobs because of cormorant predation on the region's walleye fishery. The agency hypothesized that increasing cormorant numbers were directly responsible for a large decline in the number of nonresident fishing licenses sold between 1990 and 2005. The agency then projected losses in revenue due to declines in money that would have otherwise been spent by anglers on food, lodging, gas, and supplies. Those hypothetical losses were further projected in order to estimate indirect impacts on other sectors of the economy that would arise from local businesses making fewer purchases of goods and services from other industries. The purportedly far-reaching impact of cormorants led the agency to consider an annual budget of $3.5 million for cormorant control at Oneida Lake to be cost-effective.[17]

Many factors could have influenced the declining number of nonresident anglers. But besides cormorants, the only factors the agency considered were the higher costs of fishing licenses and a downward trend in outdoor recreation in general, both of which were dismissed as unlikely causes. But significant changes in the nation's economy occurring during the same period may have affected people's expenditures on recreational activities. Additionally, a national survey conducted by the USFWS reported interest in fishing peaked in 1991, and then was followed by a declining trend nationally, with about a 16 percent reduction in the number of anglers documented between 1991 and 2006. Ironically, the importance of these potential influences was highlighted by none other than the wife of the leader of the Little Galloo slaughter gang. In a story on cormorants published by *Smithsonian* in 2003, she told the reporter that her husband "thinks it's all the cormorants' fault because that's what he sees. . . . It's not just that, of course. It's the cost of gasoline. It's that Canadians don't come here anymore because of the exchange rate. It's that people aren't coming because of publicity about the cormorants. And do you know what? Young folks just aren't fishing anymore. They don't have time to fish! Soccer practice, piano lessons, play practice. My own grandchildren don't have time to fish."[18]

11

Looking North to Canada

Limitations to Management beyond the 49th Parallel

AS THE TWENTIETH century came to a close, the cormorant problem gained momentum in Canada as well as the United States. Increasing numbers of birds and conflicts revolving around fish, trees, wading birds, and a host of other matters were reported from Alberta to Nova Scotia. Similarly, the opportunity to stir up controversy and showcase a public enemy was not missed by the media. Biased and negative press much like that in the United States was a common occurrence; more than half the studies included in the analysis of cormorant media coverage in the Great Lakes area were published in Canadian newspapers. Headlines such as "Smelly cormorants overtake old Hillsborough piers," "Thirty thousand cormorants destroying lakeside park," and "Cormorants running amok" give some idea of the cormorant's portrayal as perpetrator of environmental crime. Many studies were launched in Canadian waters to examine aspects of the bird's interaction with and impact on the environment. And as in the United States, fear and loathing spread rapidly among the public, which made demands for local and regional management.[1]

174

Tree-nesting cormorant

But in terms of cormorant control, Canada thus far stands apart from its southern neighbor. There are some superficial similarities: a site-specific approach to resolving conflicts, and actions by provincial and federal agencies as well as private landowners, which can include First Nations communities and conservation authorities. But over the last fifteen years, far fewer cormorants have been killed legally in Canada than in the United States, and government-run programs have destroyed just a small fraction of the total known number of birds killed across the continent. Numbers taken legally by private landowners are not tracked, but undoubtedly are not comparable to those taken under legal means in the United States.

Since the late 1990s, when significant cormorant management got under way in the United States, government agencies in Canada have launched culling programs at only a handful of locations. The first project was initiated by Ontario Parks to protect island biodiversity at High Bluff Island in northern Lake Ontario, part of Presqu'ile Provincial Park. Egg oiling and nest destruction began in 2003, and culling was initiated in 2004. As the project expanded, egg oiling was initiated at the park's nearby Gull Island. Egg oiling and nest destruction occurred through 2007, eliminating more than 24,566 nests on the two islands. Culling, which was limited to High Bluff Island, continued through 2006, resulting in 10,824 cormorants shot. Concurrently, management was initiated at the remote Lac La Biche area of northern Alberta by the Alberta Ministry of Environment and Sustainable Resource Development. The goal was to restore the local walleye population and the fish community in general. At multiple colony sites, egg oiling began in 2003, and culling in 2005. During the nine breeding seasons from the project's inception to 2011, these efforts destroyed approximately 11,132 birds, and eggs in more than 23,778 nests. No birds were culled in 2012. Next to be launched was a culling-only program on Lake Erie's Middle Island, part of Point Pelee National Park in southern Ontario. Initiated by Parks Canada in 2008 because of concerns about biodiversity impacts from nesting cormorants, the program had culled 8,440 birds by 2011, with one more year of shooting scheduled for 2012. Finally, culling was again initiated in Quebec in 2012, the first such program in the province since the large-scale efforts undertaken in the 1990s in the St. Lawrence estuary. This effort targeted birds at Lac St. Pierre, an enlargement of the St. Lawrence River, and was undertaken by the Ministère des Ressources Naturelles et de la Faune. In its first and thus far only year, 561 cormorants were culled. Additionally, egg oiling had been launched at the site in 2002, and 4,246 nests had been oiled by 2011. The reasons for the control included concern over declines in yellow perch and the desire to prevent cormorants from moving to other areas.[2]

In addition to these culling programs, the Ontario Ministry of Natural Resources initiated an experimental egg-oiling program in 2002 along the Lake Huron coast in the North Channel of Georgian Bay. The agency oiled eggs through 2005 at multiple colonies to determine whether cormorants affect coastal fish abundance. A total of 108,416 eggs were reported oiled, but that number included eggs that had been oiled repeatedly, and the real number oiled may have been considerably smaller.[3]

Finally, management at one additional site, the Leslie Street Spit in Toronto, Ontario, also known as Tommy Thompson Park, is especially noteworthy. Its significance, however, lies not in the number of birds killed or nests destroyed, but in the approach taken to resolving concerns about cormorants. The spit, a three-mile-long peninsula that juts into Lake Ontario, is a human-created headland where natural processes dominate. The site provides habitat for globally significant numbers of nesting colonial waterbirds, including some of the largest North American colonies of ring-billed gulls, double-crested cormorants, and black-crowned night-herons. All the more remarkable is that the spit lies just east of downtown Toronto, is visited by more than 250,000 people a year, and is considered an "urban wilderness." As cormorant numbers increased there, and concern arose over the loss of trees and potential impacts on other birds, a unique approach to managing the site developed. The highly visible cormorant colony, surrounded by a sea of people, has not been subjected to culling, nor have efforts been undertaken to reduce its size. Instead, the Toronto and Region Conservation Authority has elected to maintain the thriving colony, and is attempting to manage cormorant-induced changes and conflicts with human desires through other means. These include limiting and deterring cormorant expansion to more sensitive areas, enhancing ground-nesting opportunities for cormorants, restoring habitat, and increasing public awareness and appreciation of cormorants and other colonial waterbirds at the park.[4]

In total, since the first depredation order was established in the United States, approximately 30,000 cormorants have been killed in Canada by provincial and federal agencies. During the same period, about five times as many were killed in the United States under the Public Resource Depredation Order alone. If killings under the Aquaculture Depredation Order are considered, which includes those by private citizens and governmental agencies, the US figure is more than ten times the Canadian one.

Comparing the extent of cormorant management in the two countries is more interesting when one takes into account that Canada is home to far more breeding cormorants than the United States. The Prairie Provinces, southern Ontario, Quebec, and Atlantic Canada comprise the bulk of the cormorant's continental breeding range. Exact numbers are not available for all locations, but population estimates before widespread control got under way indicated that more than 60 percent of the Interior and Northeast Atlantic populations occurred in Canada. The true proportion may have been closer to 70 percent, because Manitoba-wide numbers were not available for the Interior estimate and included only birds at Lake Winnipegosis.[5]

Canadian Differences

How is it, then, that cormorants are managed in so few locations and in such smaller numbers on Canadian soil? While Canada has a smaller and less dense human population, the abundance of birds there leads to conflicts like those in the United States, especially in areas more heavily populated or used by humans. What prevents the Canadian government from engaging in intense cormorant management? Do Canadians have greater tolerance for this species than Americans? Many factors likely contribute to this more hands-off approach.

Cormorant Distribution and Aquaculture

One reason the geographic extent of cormorant management is more limited in Canada relates to the bird's distribution. Cormorants do not overwinter in Canada in significant numbers, and there is a six-month period when human conflicts with the birds do not occur. And while aquaculture is well developed in Canada, and conflicts with cormorants over human-raised fish are documented, nothing there is even remotely similar to the Delta region of Mississippi. There are numerous examples of human-altered fish communities potentially benefiting cormorants, but no major portion of the bird's breeding range has been transformed into an artificially stocked, easily accessible, high-quality feeding area.

Provincial Management Authority

Another important reason for Canada's limited management of cormorants is related, ironically, to the birds' lack of federal protection there. As previously described, the Canadian government has not added cormorants to its federally protected bird list identified in the Migratory Bird Convention Act. Instead, cormorants are protected by acts of the provincial legislatures, which also have responsibility for managing the species.

Management practices and attitudes toward cormorants have varied among the provinces. Authorized killing of the birds is limited through specific local permits, hunting seasons, and varmint licenses (in a few locations), or under circumstances specified by provincial policy. In turn, provincial policy is influenced by government policy, stakeholders, and agency attitudes.

In Nova Scotia, the species is fully protected under the Nova Scotia Wildlife Act, and no population control programs have been undertaken. Site-specific problems involving commercial fisheries and stocked fish ponds that can't be otherwise addressed are resolved by issuing permits to "bona fide" commercial fishermen and pond owners, but birds are not allowed to be shot at nesting colonies or on inland waters except at stocked ponds. For concerns raised by recreational fisheries, the Nova Scotia Department of Natural Resources considers cormorants "part of the endemic predatory component of freshwater, estuarine and coastal marine ecosystems of the Province of Nova Scotia." It therefore recommends that no permits be issued on inland waters, where most complaints concern salmon, unless extraordinary circumstances prevail. It suggests that cormorant predation be included in salmon population modeling, and that the timing and location of smolt releases be altered to minimize predation. As for cormorants' effects on vegetation, the department emphasizes that cormorants are part of the natural system and Nova Scotia's wildlife heritage, and that cormorant-induced habitat changes appear beneficial to at least some species. Therefore, no control or harassment is allowed on provincially owned sites. On private lands, owners are allowed to protect private property, but such actions are the owner's responsibility, and a brochure has been issued with guidelines.[6]

In Ontario, cormorants are protected by Ontario's Fish and Wildlife Conservation Act, and the province has specific policies for cormorants on both Crown (public) and private lands. The policy for cormorants on Crown land, published in the *Ontario Environmental Registry* in 1998, specified that "control of cormorant numbers should only be considered in specific local areas if birds are found to be having significant negative, ecological impacts on specific habitats or other species." Further, it directs the Ontario Ministry of Natural Resources "to confirm reasonable evidence of significant negative effects on natural resource values at the target site prior to undertaking control activities."[7] The policy is largely a product of the current "green" government in Ontario, which makes biodiversity conservation a government-wide priority. Notably, cormorants are not managed in order to leave more fish for humans, a practice considered inconsistent with a priority of biodiversity conservation. Although Ontario's Fish and Wildlife Conservation Act does not apply to provincial or federal agencies on Crown lands, meaning that a federal agency can manage cormorants on federal public lands without provincial authorization, federal agencies generally try to comply with the spirit of provincial policy and regulations. The relative stability of Ontario's government (currently in its third term) and its consistent priorities (such as biodiversity protection) have enabled the province to maintain this position since the policy was established.

On private land, the Ontario Fish and Wildlife Conservation Act allows landowners to protect their property from wildlife damage, and landowners or

their immediate family members can harass or kill wildlife damaging or about
to damage their property. The landowner doesn't require authorization in order to harass or kill the wildlife, but does require provincial authorization to
destroy nests or eggs. Landowners can hire an agent to act on their behalf, but
the agent must be authorized by the Ministry of Natural Resources. Landowners are authorized to act only to prevent further damage and must not cause
unnecessary suffering.

In Alberta and Quebec, less conservative policies have developed. In
Quebec, cormorants are protected under the province's Wildlife Conservation
and Management Law. While cormorant policy is not formally documented,
the control programs that have developed since the 1970s indicate that the
province manages cormorants to address both biodiversity and potential or
indirect impacts to fisheries. In 2006, the Ministère des Ressources Naturelles
et de la Faune prepared a fact sheet that noted: "Various factors (ecological,
environmental, social, economic, etc.) must be taken into account when decid-
ing whether or not to initiate control measures." Lac St. Pierre was presented
as an example of a site where multiple factors influence decisions about cormorant management. Declines in fish began before cormorants colonized the
area, indicating that other factors were at work. It was nonetheless believed
that cormorants were contributing to the lack of perch recovery. Currently, a
five-year moratorium has been placed on perch fishing, and the ministry wants
to work on all factors affecting the perch population. This suggests that the
province will manage cormorants wherever some evidence demonstrates that
cormorants are affecting resources.[8]

In Alberta, where cormorants are protected under the Alberta Wildlife
Act, a policy to manage them specifically to protect fisheries from predation
has developed. In 2002, a bill was introduced for the active management of
fish-eating waterbirds. Although the bill has not moved forward, Alberta Sustainable Resource Development (ASRD) began controlling cormorants in
2003 in response to public concerns expressed over the impact of cormorants
on Lac La Biche fisheries and in order to reach objectives for fish populations
there.[9]

In Manitoba and Saskatchewan, despite the fact that each province supports very large numbers of breeding cormorants, experiences strong political pressure from fishermen to control the birds, and a great deal of illegal
killing, the provincial authorities have not agreed to control cormorants. For
these provinces, I could not obtain detailed information on formal policy, so

the reasons for these decisions are not clear; nevertheless, they demonstrate how provinces vary in their approaches and their responses to public pressure.

These examples indicate that provincial-level authority, along with regionally distinct attitudes toward cormorants and the resources they employ, has prevented a single approach to cormorants from developing across Canada. These provincial differences suggest that Canada continues to be a long way from any national approach to cormorant management.

A Lack of Federal Management for Nuisance Wildlife

A third factor in Canada's less aggressive management approach is the lack of a Canadian counterpart to the USDA's Wildlife Services. There is no federal agency in Canada devoted to or responsible for managing "nuisance" wildlife. Nor are federal funds allocated to supplement provincial management projects that do so. While the Canadian Wildlife Service issues permits to oil eggs or kill federally protected birds considered a "nuisance," such as Canada geese, the agency does not do the control work itself. Instead, permits are typically issued to landowners, who then have to perform the control work or hire a private company to do it. In the few instances of recent cormorant control, management plans and culling have been carried out by both provincial and federal natural resource agencies. Rather than being dedicated to agricultural development and protection, these agencies typically have a broad focus on stewardship, biodiversity conservation, sustainable ecosystems, and natural resource use.

Additionally, neither federal nor provincial agencies have any responsibility to manage cormorants, and certainly no conflict has arisen between agencies for management authority. Furthermore, no federal agency with any authority for wildlife control has issued a position statement on cormorants, drafted an initiative for cormorant management, or otherwise encouraged provinces to develop large-scale or flyway-level management plans. This is not to say that provinces don't experience pressure from stakeholders to manage cormorants. But provincial agencies don't experience additional pressure from federal agencies representing specific interests.

Because there is no single leading agency advocating or implementing large-scale cormorant management, no framework has been established or precedent set for coordinated or geographically widespread cormorant control. In the absence of such an influence, cormorant management has

developed more slowly and has been more conservative in Canada than in the
United States.

The Power of Opposition

Perhaps the most distinct and important influence limiting cormorant management in Canada has been the development of stakeholders in Ontario who have
strongly challenged cormorant management. Six groups consisting mostly of
animal-rights and environmental activists, along with a handful of especially
determined individuals, have shown a remarkable tenacity and resourcefulness
in opposing cormorant control: the Peaceful Parks Coalition, the Animal Alliance of Canada, Born Free USA (most of its activity regarding cormorants is
in Canada), Zoocheck Canada, Earthroots, and the Toronto-based Cormorant
Defenders International, a coalition of Canadian and US groups with a mission to educate the public, dispel misinformation, and advocate on behalf of
cormorants.

In their battle, the activists have been highly strategic and were involved
in Ontario's cormorant problem as early as 2002, when the first management
plan for cormorants was developed for Presqu'ile Provincial Park. Much of
that effort to oppose cormorant control was spearheaded by Peaceful Parks.
This organization, along with the Animal Protection Institute, wrote letters
of opposition and immediately began efforts to inform the public about the
issue. When Ontario Parks made the decision to cull birds at Presqu'ile, an
intensified campaign was launched to raise public awareness and challenge the
government's plan. Shortly before the cull began, Peaceful Parks hired a small
plane to fly over Queen's Park with a simple message for Ontario premier Dalton McGuinty: "Save The Double-crested Cormorants."[10]

When culling was initiated in May 2004 at High Bluff Island, the battle
got under way in earnest. The activists recognized that showing people what
a cull actually looks like could lead many to question the necessity and ethics
of its undertaking. Therefore, the initial focus was to identify and publicize the
humane issues arising from culling, egg oiling, and nest destruction. The activists employed myriad tactics aimed at increasing public awareness and directly
interrupting control activities. Peaceful Parks handed out flyers to park visitors
about the cull and organized canoe trips to the island for the media. During the
cull, members of the organization floated alongside the island and bore witness
to the killing. The Ontario Ministry of Natural Resources would not shoot

birds if canoeists were in the firing line, and thus activists attempted to be in the line of fire as often as possible. In this first year, staff and volunteers from Peaceful Parks, Zoocheck, and Earthroots spent days on the water, bearing witness to and documenting the cull, while other individuals ran interference. Some arrests were made, intensifying the issue and raising public awareness. Peaceful Parks issued numerous press releases describing the cull, critiquing the government, and inviting the public to come and see the cull for itself.[11]

In 2005, activists continued these activities; enlisted a team of veterinarians to retrieve, dispatch, and, when possible, save injured birds; and began obtaining video footage of the cull. The footage was later incorporated into a twenty-five-minute documentary produced in 2007 by Cormorant Defenders International. *Cormorants in the Great Lakes: Dispelling the Myths* vividly demonstrates that for many birds, a cull provides anything but the antiseptic and humane death such an operation hopes to achieve. Shot cormorants are shown limping across the island, dragging ruined wings, floundering in the water, and dying in the boats of activists who have retrieved them from the water or the island's shore. Close-ups of grisly wounds on still-living birds emphasize that the cull is not without some degree of horror. The camera follows birds that can no longer fly, tend to their nests, or feed themselves; they simply wander about the island, drift or drown in the water, or huddle on the shore. The documentary describes observations of some wounded birds lingering for as long as five to seven days and records pleas from activists to park personnel and shooters to retrieve the birds, which fall on apparently deaf ears. Powerful commentary is provided by Mary Richardson, a veterinarian shown inspecting shot birds in the cull's aftermath. The sound of waves and the shrill cries of gulls can be heard in the background as she addresses the inherent cruelty of such an operation. In one particularly moving scene, she examines a still-living bird that has been shot through the abdomen; the bullet has exited the other side of the bird's body and shattered its leg. As she traces the path of the bullet and discovers the extent of the wounds, the bird's chest continues to rise and fall, a small movement suggesting volumes. Describing the bird's injuries and realizing it is beyond all repair, Richardson cannot keep the emotion from her voice.

Before the release of the documentary, activists began using the video footage to publicize cormorant management as a significant animal-cruelty issue by distributing a four minute clip titled "Carnage at Presqu'ile." In addition to posting it on YouTube, activists distributed it at the park gate to park visitors, to homes in Brighton (the community next to Presqu'ile), to people on the street, and at big concerts in Toronto.

In addition to documenting the direct effects on birds being culled, the activists recorded observations on potential casualties of the cull, which brought them their first major victory. In 2005, activists filmed parental feeding behavior in the colony while culling was occurring. Because orphaned young birds suffer a particularly cruel fate, often dying of starvation, exposure to the elements, or predation, the Ontario Ministry of Natural Resources had committed not to shoot birds with young. As soon as activists observed feeding behavior, they notified the ministry and Ontario Parks that chicks were present. But despite their reports, the ministry continued culling activities. In response, Peaceful Parks issued a press release stating that it had notified the ministry and Ontario Parks of its observations of feeding behavior during a cull. Other groups issued similar press releases. Two days after the first press release was issued, the ministry halted the cull, a premature end that spared an estimated 3,363 birds from being shot. Later, a report prepared by the Presqu'ile Double-crested Cormorant Management Scientific Review Committee for the ministry acknowledged the significance of the 2005 protester activities. The committee noted that protesters had forced the park to alter its management plan for the year, resulting in a significantly smaller number of birds being killed than what was targeted.[12]

In 2006, governmental agencies made a major effort to exclude observers from interfering with and documenting the cull. An exclusion zone was established for all boaters, along with a strong police presence on the water during times of shooting. More arrests were made, but the exclusion zone was ultimately deemed illegal, and charges against the activists were dropped. So ended the final year of the culling program. The egg-oiling and nest-destruction programs concluded in the following year.[13]

In 2008, the Middle Island cull was initiated, and the activists again provided a constant presence to observe and document activities. But events around this cull proceeded somewhat differently than those at High Bluff Island. The culling protocol, which reflected lessons learned during the High Bluff Island cull, was modified to minimize humane issues documented by the activists. It is also the only culling effort anywhere for double-crested cormorants that has collected specific information to document wounding rates. These data indicate that extra precautions help minimize suffering, but it still

Parents feeding nestlings

takes more than one shot to kill some birds, and a small number of the injured can't be dispatched. For this cull, a legal exclusion zone was established around the island, and activists were not able to observe culling activity directly. An exception was made to allow Cormorant Defenders International to watch from a boat fifty yards offshore, but activists had no visual access to the island's interior. The activists continue to document their observations and focus attention on related animal welfare issues.[14]

While attention to humane issues has been a major part of the strategy to challenge cormorant control, activists realized that it was just as important for the public to question whether the cull was really necessary and to consider the possibility that the birds might not need to die. Because governmental agencies presented ecological arguments for their actions, alleging that protecting habitat from cormorants was a biological necessity, cormorant management could be perceived as a necessary evil. Activists recognized that if the humane issues were not tied to the underlying rationale for management, the public could be lulled into acceptance by improved culling protocols, and culling would likely continue, just as it had at Middle Island. Therefore, activists have prepared detailed critical analyses of the specific rationale underlying management at both locations. In an effort to educate themselves about waterbird science and policy, they have attended many scientific meetings and presentations, drawing on the knowledge of scientists, ornithologists, and naturalists across North America. In addition, they have held their own workshops and symposia, which featured naturalists, veterinarians, biologists, lawyers, policy experts, and other skilled professionals.

Over time, they have become formidable opponents of those in favor of cormorant management, and have articulated several well-reasoned ideas that are relevant not just for these locations but also, more broadly, for the general issue of cormorant impacts to islands and biodiversity. They have raised important questions that revolve around how concepts of biodiversity, ecological integrity, and overabundance are used. Central to their argument is the premise that, rather than the cormorant's current abundance and distribution, the cormorant's diminished presence on islands in the past was actually the indicator of poor ecological conditions. Human persecution and environmental contaminants, rather than natural stressors, were responsible for the cormorant's decline; thus, conditions that developed in their absence were not fully representative of the natural community.

Activists have worked to make their findings readily available to government agencies and the public. Full critiques, analyses, reviews, and discussion of management are posted on the websites of Cormorant Defenders International and the Peaceful Parks Coalition. They have informed the media of alternative opinions and approaches to cormorants and have generated substantial negative press for managing agencies. Information is also dispersed during demonstrations, workshops and symposia, and when activists attend scientific meetings. Further outreach efforts have employed diverse tactics, ranging from a billboard campaign about cormorants in Windsor, Ontario, the largest city near the Middle Island cull, to street theater performances in several areas.

In addition, activists have worked through other channels to further influence cormorant management. They have been able to make highly effective use of Ontario's substantial processes and procedures that must be completed before cormorant management can be pursued. Because of these requirements, they were able to prevent or delay management actions and doggedly challenge just about all aspects of Ontario's two culls. They also launched extensive letter-writing campaigns that sent thousands of opposition letters to provincial and federal governments. And they have pursued lawsuits against governmental agencies.

How influential has the opposition been? Between 2006 and 2009, Ontario Parks proposed additional lethal management for cormorants at Presqu'ile Provincial Park and identified cormorant control as a potential strategy for East Sister Island, a provincial park and nature reserve in western Lake Erie. The plans were immediately challenged by activists, and as of 2012 no substantial lethal management has occurred at Presqu'ile since the major efforts ended in 2007, nor has management been initiated at East Sister Island. While the extent to which activist opposition influenced these decisions is not known, the activists are acknowledged by natural resource agency professionals as invested and well-organized stakeholders opposing the development and implementation of aggressive cormorant management in Ontario. Additionally, the more humane culling protocol used at Middle Island represents an important achievement for the activists, since it was largely through their efforts that serious humane issues were recognized.[15]

At Toronto's Tommy Thompson Park, activists have clearly influenced management policy. To consider appropriate cormorant management strategies for the park, the Toronto and Region Conservation Authority established an advisory group that included federal and provincial agencies, citizen advocacy groups, and animal-rights and environmental activists. The agency gave consideration to all stakeholders and defined a goal "to determine an effective, humane and acceptable management approach to cormorants at TTP." The advisory group emphasized a humane approach as part of the selection criteria for appropriate management strategies, an emphasis that reflects the Toronto and Region Conservation Authority's position not to cull cormorants. Furthermore, agency actions to promote an awareness and appreciation of cormorants reflect stakeholder values that consider large numbers of cormorants natural and representative of a healthy ecosystem.[16]

In conclusion, it is important to note that no comparable opposition has developed anywhere in the United States. Nor has such a diverse group of stakeholders been invited into the management process for any of the plans that have been developed there.

The Importance of Location

Comparing the public response to culls in Ontario and Alberta suggests regional factors have been important determinants for support or opposition to cormorant management in Canada. Southern Ontario, the only area where cormorant management has received substantial opposition, is the most densely populated region of Canada. Tommy Thompson Park is just minutes from downtown Toronto. High Bluff Island, only about a mile offshore, is minutes away from the city of Brighton, and Presqu'ile Provincial Park is a half hour's drive from the well-developed city of Peterborough. Middle Island is a more remote location, but Point Pelee National Park is only about a half hour's drive from Windsor, a medium-sized Canadian city with a substantial tourism industry. These larger metropolitan areas are more likely to include animal-rights and environmental activists than less populated or remote, rural areas. There are at least ten animal-rights groups or organizations in the Toronto area, others in nearby Hamilton and Niagara, and one in Windsor. These groups frequently pay close attention to governmental programs affecting animals and wildlife in local areas, and several have been involved in opposing cormorant management.

By contrast, the province of Alberta is much less densely populated than Ontario, and most of the population resides in the southern portion of the province, in the Calgary-Edmonton corridor. The Lac La Biche region, located in the northern portion of the province, lies some two hours north of Edmonton and approximately four and a half hours from Calgary. The cormorant colonies being managed there occur on much smaller and accessible lakes than the Ontario sites, but the remote northern location by itself presents a challenge simply because of its distance from any major urban concentration. Such a remote area is unlikely to be populated by animal-rights groups or other activists interested in protecting cormorants. When cormorant management at Lac La Biche first got under way, some opposition was mounted by the Alberta

Wilderness Association, which viewed the program as an attempt to scape-goat cormorants for poor fisheries practices. But scarce resources limited the group's focus to key wilderness conservation matters in northeastern Alberta. Thus far, no other group has stepped in to pose a challenge.[17]

The specific goals of the management plans and the interests of stake-holders are also related to this regional aspect. At Lac La Biche, the goal of management was straightforward: reduce cormorants in an effort to restore the fish community to a composition preferred by humans. Fishing is a main-stay of the area's culture and has historically been very important. Lac La Bi-che is the region's biggest lake district, and the communities in the area have a high proportion of fishers. Additionally, people from outside the area go there to fish. Fort McMurray, an industrial area constituting one of Canada's major oil production hubs, is about two hours north, and many of this city's workers unwind in the Lac La Biche area. The community thus has a strong interest in the composition and structure of the fish population. The Ministry of Alberta Environment and Sustainable Resource Development approached cormorant management in terms of community needs and within the context of a com-plete restoration program. Measures of the program's success were defined as populations of walleye, pike, perch and whitefish able to support more human harvest. As discussed earlier, Lac La Biche is an area where human overfish-ing resulted in a dramatically altered fish species composition. The fish com-munity went from one dominated by large predator species, such as walleye and pike, to one dominated by small forage species, such as perch, cisco, and shiners. Residents accepted that large-scale ecosystem changes were necessary to restore some of the social benefits derived from the lake—namely, fishing opportunity—and cormorant management in particular has been strongly en-couraged here. Considered in this regional context, it comes as no surprise that there has been little opposition to cormorant management at Lac La Biche.[18]

In comparison, both Ontario culling programs identified goals of restor-ing or protecting vegetation and island diversity by reducing cormorant num-bers. Those goals raise some controversial ecological and conservation issues that are discussed in detail in later chapters, but a preview of them includes the following. First, High Bluff and Middle Islands are officially recognized for their importance to nesting colonial waterbirds, including cormorants. High Bluff Island, in conjunction with Gull Island, forms a breeding-bird sanctuary, and Middle Island is part of an archipelago important for breeding and mi-gratory birds. Second, protection and recognition of these sites as important island ecosystems has been fairly recent; throughout much of the twentieth century, people frequently used—and disturbed—Middle Island. Additionally, High Bluff Island was connected to Presqu'ile Point and the mainland by a narrow causeway until the mid-1950s, becoming an island again only when a hurricane swept away the causeway. Much of the vegetation on both islands is nonnative. Third, islands and bird populations are dynamic, and the species composition on islands often changes over time. This combination of factors led many stakeholders in Ontario to challenge the goals of the culling programs. Arguing that change as a result of cormorant activity is not necessarily an indicator of a degraded ecosystem, they suggested that parks should be promoted as evolving and dynamic ecosystems rather than managed as "green museums frozen in time."[19]

At Tommy Thompson Park, habitat change due to cormorants was more readily viewed in the context of natural change, and the history of the site provides an important background for the management that has occurred there thus far. The park resulted from construction that began in 1959 to enclose a basin and create Toronto's Outer Harbor. The main construction continued for two and a half decades and included the creation of peninsulas from dredged silt and sand, which fostered rapid ecological succession. The site eventually began to support provincially and nationally rare plant species, regionally rare habitat, and a significant stopover site for migrating birds. These developments led to its designation as an Ecologically Sensitive Area in the 1980s, which was updated in 1993 to recognize the spit as the only breeding habitat for cormorants in the jurisdiction of the Toronto and Region Conservation Authority. In 2000, the site was designated an Important Bird Area by Bird Life International, an organization with a mandate to conserve important sites for all bird species worldwide. Thus, many stakeholders considered the presence of cormorants and their interactions with other colonial waterbird species and vegetation as part of the manmade site's natural development.[20]

In considering cormorant management across Canada and the United States, southern Ontario presents a unique picture. It is the only North American location where a range of resourceful individuals and groups have come together to challenge management policies effectively. Ontario's policy requirements for public consultation have been partly responsible for that success. The sites where management has been challenged are all parks, and public consultation for management at parks is critical. What is more, they share similar reasons for initiating cormorant management, which inspire debate over what constitutes wilderness and ecological integrity. Compared to sites managed specifically to benefit people through fishery allocations, sites such as provincial and federal parks may have considerably more support for a hands-off approach. Finally, in contrast to US practice, Ontario's policy requires evidence of impacts before cormorant management can begin. Thus Ontario provides the exception to "business as usual" and is the only location where important strides have been made toward a more humane and tolerant approach for cormorants.

PART IV

The Science, Management, and Ethics of Today

Review and Critique

12

Untangling the Mysteries
between Predator and Prey

Nature loves to hide.

—Heraclitus of Ephesus, 6th century BC

AT THE HEART of the cormorant's story is the extent to which its current treatment is (or is not) based on sound science, especially relative to its management for fisheries. Since the late 1990s, the vast majority of birds and eggs destroyed have been targeted specifically to protect fish for humans. But the extent to which cormorants harm human interests by consuming fish has not been easy to determine. While theory, observation, and experimentation have established the potential for predators to affect prey populations, there are many situations in which predators do not control the abundance of their prey, and sometimes even have effects contrary to expectations. How a cormorant affects its prey at any given time and place ultimately depends on many highly variable factors and interconnected processes. Lying deep in question is the gray area of circumstances in which predation will lead to effects on fish populations that are substantial enough to consider management. In this regard, uncertainty has become a hallmark of the cormorant-fisheries conflict, and only rigorous scientific studies that examine how cormorants interact in ecosystems can determine whether negative impacts are occurring.

A pursuit dive

Some Considerations regarding Impacts

To date, no study has demonstrated that cormorants pose a significant threat to the survival of healthy fish populations in natural systems. Limited data for prey depletion around cormorant colonies exists, but there is no evidence that double-crested or other cormorant species have ever depleted fish populations severely enough to pose an extinction risk. There are examples where predation on already threatened fish species may be enough to push local populations over the edge, such as endangered runs of salmon in North America or threatened whitefish in the United Kingdom. But there are no known cases of cormorants being the root cause of a species' threatened status. Thus, the need to manage cormorants does not arise because they threaten the survival of fish populations, except where populations are already compromised.

But natural resource agencies consider the maintenance of fisheries for human use to be a conservation goal, too, and top-level predators can certainly affect fishery yields. These considerations define the cormorant-fisheries conflict as an issue of resource allocation rather than one of ecosystem health. The extent of the problem is determined by how much of the resource humans claim for themselves and how much they are willing to share with cormorants. Thus, decisions to manage cormorants to meet fishery objectives are based on politics rather than ecology. But even these sorts of political decisions need to be informed by knowledge of underlying ecological interactions so that management effects can be predicted.[1]

It is important to realize that impacts to fish populations are not quite the same as impacts to fisheries. A fishery is defined not only by a particular species of fish and its population size but also by the humans involved, the fishing methods used, and the area of water being managed for harvest by humans. It is therefore possible for predation to lower human harvests but to be unimportant to, or even improve, the overall health of the fish population. Conversely, it is possible for predation to depress a fish population but be undetectable by humans if the fishing remains good. Therefore, scientific studies must define how an impact on the fish population translates into an impact on human priorities. Typically, an impact becomes significant enough to trigger management action when declines in the human catch and harvest then lead to economic losses from reduced commercial sales or diminished recreational use of an area dependent on a fishery.[2]

Data Requirements for Assessing Impacts

While the impact to fisheries from cormorants has been a topic of much discussion, there is no overall agreement on the requirements for assessing the
influence of predation. In an effort to bring fisheries and waterbird scientists
together to examine approaches to this complex task, a workshop was held in
November 2000 in Plymouth, Massachusetts, and attended by scientists from
Europe and North America who study cormorant-fishery interactions. Participants were asked to reach consensus on methods for measuring the cormorant
diet and requirements for quantifying impacts to fisheries. Results from the
workshop, which were published in 2003, provide some potential guidelines for
impact studies. Additionally, a significant body of work on assessing cormorant impacts at Oneida Lake addresses data requirements. More broadly, work
by the theoretical ecologist Peter Yodzis, who focused largely on the ecology
of food webs, discusses the complexity of and requirements for understanding
predator-prey interactions. Below, the data requirements for assessing impacts
as gleaned from these sources are outlined step-by-step.[3]

Fish Population Size and Trends

An estimate of fish population size enables researchers to estimate the proportion of fish being consumed by cormorants, and the magnitude of cormorant
predation. But this is a difficult estimate to obtain, and most studies are unable
to provide it. Alternatively, population indicators obtained through fish surveys are typically used to identify trends; they are most useful when compiled
over a series of years. That kind of information is essential for understanding
natural population variation, which helps distinguish real declines from natural fluctuations in population size.

The Spatial and Temporal Distribution of Fish

Because fish move throughout bodies of water during the year, knowledge of
their spatial and temporal distribution is necessary. For instance, yellow perch
often figure prominently in the cormorant diet when fish are spawning and
using shallower waters, but once perch disperse, they are often consumed
less frequently, which buffers the overall impact to the population. In large
or coastal waters, fish distribution and movements over a wide area may help

dilute predation effects, while in smaller, inland waters and bays, effects may be more pronounced.[4]

Fish Population Dynamics

Knowledge of the age composition, mortality, and survival and growth rates of a fish species is key for understanding impacts on a fish population. An understanding of how the population is structured (age composition) is necessary for assessments of fish stock and recruitment, while information on mortality, survival, and growth rates is needed in order to predict a fishery's future yield. Fishery recruitment, a complex topic, summarizes the interplay of factors that determine when a fish is old enough to be caught and harvested, and how many fish will reach that age in a particular time period. In general, young fish have very high mortality rates, and most do not survive to become part of the fishery (that is, they do not "recruit"); but as surviving fish get older, mortality rates decline. Once year-class strength (that is, all fish hatched in a particular year) is established, the abundance of that particular year class and recruitment to the fishery can be determined.

Site-Specific Information on Fish

At any given time and place, a wide variety of dynamic factors operate on fish populations and determine the extent to which cormorants will influence abundance and recruitment. For example, shifts in predator or competitor populations can lead to increased predation on desirable sport fish or, conversely, buffer predation on species that humans value. Variation in local weather patterns and temperature can affect the timing of fish spawning and cormorant migration, making fish more or less vulnerable to predation. Changes in aquatic vegetation over the course of a season may also alter the amount of safe habitat for fish. Therefore, if the rationale for managing cormorants is to reduce predation pressure on a particular fish population, site-specific monitoring of fish and the factors affecting them must be ongoing.[5]

Cormorant Diet Assessment

As early as 1894, a paper was published in the *Auk* by George Mackay, who was curious about the diet of cormorants roosting at a long-time haunt, Cormorant

Rocks at Seconnet Point, Rhode Island. On a single day in April 1892, Mackay shot birds to examine what was in their stomachs, collected pellets expelled by the birds containing the indigestible parts of prey, and observed the items regurgitated by one of the birds he wounded. Those three approaches are still the most commonly used methods to examine the cormorant diet today. The following discussion of their advantages and limitations indicates the complexity involved in simply assessing what cormorants eat and highlights the need to scrutinize diet-assessment methods when evaluating impact studies.[6]

Cast pellets often contain teeth, scales, and bony structures that can be used to identify the species, size, and age of fish consumed. They can be easily collected at breeding colonies and roost sites, do not require killing birds, represent foods consumed over a period of time, and can provide a rough index of the cormorant diet. But the remains recovered have undergone the entire digestive process in the animals that consumed them. The otolith, a small bony structure of the inner ear, is a diagnostic structure of fish frequently used to examine the cormorant diet, but during digestion, the otoliths of small or young fish can disappear completely, and those of larger fish may change in size and shape, rendering them useless for identification or leading to underestimates of fish size and age. To minimize this bias, some investigators have developed an otolith classification scheme according to degree of erosion, but others caution that eroded otoliths can appear pristine, and thus attempts to sort out eroded from uneroded ones may be misleading. Another potential bias is the assumption that one pellet represents a cormorant's daily food intake. Recent unpublished studies with captive double-crested cormorants indicated that average pellet production time ranged from 1.2 to 1.8 days but could take up to a week, and varied according to fish morphology. Similarly, for the closely related European shag, work with wild birds indicated that a pellet was produced on average every 3.7 days, and that the rate of pellet formation may be depressed in summer. A third source of bias creeps in when otoliths retrieved from pellets are actually those of fish consumed by another fish, which was then later consumed by a cormorant.[7]

Compared to pellets, analyzing the contents of food regurgitated by birds disturbed at colonies or roosts is cheaper and easier. Frequently, fresh or relatively undigested fish specimens are obtained, including small and soft-bodied prey that may be absent from or poorly represented in pellets. Specimens can be easily identified and measured for estimates of length and age. But the largest part of the sample typically obtained may be from chicks, and because adults may consume different food items, their diet may not be well represented. Additionally, this method assesses the contents of only very recent meals and provides little information on daily diversity within the diet.

Stomach contents can provide the most rigorous assessment if the sampling framework is carefully designed. Advantages include obtaining samples of fresh, undigested contents if they are obtained shortly after birds have fed, the ability to distinguish recently consumed prey from those that are partially digested, and site-specific feeding and demographic information. Drawbacks include the use of lethal or invasive methods to obtain samples, which besides being time-consuming and costly, can result in small sample sizes. If the study is not carefully planned, empty stomachs are frequently obtained, further reducing sample size. Additionally, temporal and spatial variation may be undersampled, and differences between individual versus group foraging can lead to biased estimates of diet. Finally, this method has limited usefulness for assessing daily diversity and food intake because it is not typically known whether birds are finished foraging for the day.

Cormorant Consumption Rates and Number of Birds Feeding

The predation of older rather than younger fish may have a different effect on the number recruited to or removed from the fishery. Thus, studies must take account of consumption rates that are age-specific for fish species of interest. Study design should also take into account factors influencing consumption rates, such as season, the energetic demands related to activities, and the caloric values of prey species. Additionally, accurate estimates of the number of birds feeding in the area are necessary, but because breeding populations are composed of both breeding and nonbreeding birds, the process of obtaining such estimates is not straightforward. Breeding numbers are based on the assumption that an individual nest represents one pair of birds, and nests can be easily counted. But nonbreeding birds are not tied to a discrete site, and their numbers are far more challenging to determine. Observations have provided a wide range of estimates, from 0.6 to 4.0 nonbreeding birds for each breeding pair, but most were not obtained through rigorous scientific study.[8]

Site-Specific Information on Cormorants

In 1892, diet sampling led Mackay to conclude that eels constituted a large part of the birds' food. But he was quick to point out that his results provided but a narrow window into what cormorants were eating at a particular time and place. From the myriad diet studies conducted since Mackay's time, a basic observation has emerged time and again: the cormorant diet is highly variable. Studies have documented significant differences in diet between seasons, years, reproductive stages, birds of different sexes and ages, and birds at the same lake and at different ones. This variability and partitioning may lessen impacts on fisheries and indicates that extrapolations and generalizations about what cormorants eat and how they influence fish may be misleading.[9]

Assessing Impacts
Food Webs and the Environmental Context

Well-designed studies recognize that the effects of predators on fish populations occur within a food web, and even the simplest food webs have many pathways and consumers. In an elegant review paper published in 2001 titled "Must Top Predators Be Culled for the Sake of Fisheries?" the ecologist Peter Yodzis provided a detailed discussion of the complex relations between predators and fisheries. Yodzis focused on the food web of the Benguela ecosystem on the southwest coast of Africa. The ecosystem supports an important fishery of hake, an ocean fish similar to cod. Fur seals live there, and their consumption of hake has led the South African government to consider culling their numbers. But the Benguela ecosystem food web contains at least twenty-eight million pathways, and the removal of one predator may not necessarily make a desired fish species more available to a fishery. Yodzis used a simple but powerful figure to graphically represent the maze of pathways and demonstrate how, as in a domino effect, additional interactions may occur when one predator is removed. Other predators in the system may respond to increased fish abundance, or competition between fishes may increase; it is even possible for the removal of a top predator to result in interactions that diminish fishery yields. When making management predictions, fisheries scientists approach this complexity by distilling it down to the interactions they consider most important. But how well they are able to do so depends on how well they understand the system in question.

A study published in 1995 by the fisheries and wildlife scientists Charles Madenjian and Steven Gabrey examined the role of fish-eating waterbirds, including double-crested cormorants, in the food web of western Lake Erie. For each waterbird species and for the group as a whole, total fish consumption was estimated and then compared to that of walleye, western Lake Erie's dominant fish-eating vertebrate. Results indicated that over the course of a year, waterbirds ate the equivalent of just 15 percent of the fish consumed by walleye in a single growing season, and cormorants consumed the equivalent of just 2 percent. Additionally, red-breasted mergansers, herring gulls, and ring-billed gulls, present in larger numbers than cormorants, consumed more fish than cormorants during the period examined. The study highlights the importance of considering the effect of cormorant predation in relation to that of other dominant predators, and demonstrates that the most influential predators are not necessarily the most visible ones.[10]

Work by Jennifer Doucette and her colleagues at the University of Regina, Saskatchewan (published in 2011), examined how cormorants fit into lake food webs and how this structure might be relevant for understanding impacts on fisheries. Results showed that the position occupied by cormorants in the food web was a potential determinant of impacts. The authors reported that impacts "may not be a concern in many cases because negative effects may only result from instances where the dietary niche of cormorants overlaps with, or is composed of, economically valuable fish species." In turn, the degree of overlap between cormorants and other fish eaters may be determined by the local food web structure rather than by cormorants using all available predator niches. The dynamics again indicate that understanding cormorant impacts likely requires a multispecies and food-web-based approach, especially if management targets healthy, stable food webs and not just individual species. Further, because food webs are unique, they should also be considered individually; it should not be assumed that cormorant impacts are the same everywhere.

Other environmental factors that can influence food web structure and fishery trends include changing temperature conditions and water levels because of a changing climate; the introduction of invasive species that can prey on or outcompete native fishes or change the prey base; changes in nutrient levels and lake productivity; and illegal harvests of fish. Regarding that last factor, evidence of extensive underreporting of fish harvested has been obtained from many areas where cormorants are considered a principle factor in fishery declines, including the eastern basin of Lake Ontario, the Bays de Noc (Michigan), and the Wisconsin waters of Lake Michigan. The scale of the illegal harvests can be significant. In Wisconsin, commercial fishermen admitted to underreporting harvests that were two to three times larger than reported harvests in the early 1990s. In Michigan, the Department of Natural Resources reported illegal harvests of roughly 72,000 pounds of walleye from Little Bay de Noc between 2004 and 2009. In 2013, the results of a three-year investigation of northern Minnesota lakes revealed that thousands of fish, mostly walleye of various sizes, were being illegally bought and sold, and that many other less valuable fish were being illegally dumped and wasted. Many walleye came from Leech Lake, where cormorants are intensely managed to protect this species. Although the impact this illegal activity had on fish populations is not known, the Minnesota Department of Natural Resources acknowledged that it "clearly removed fish from lakes that could have been caught legally and utilized by state anglers and tribal fishers."[11]

Modeling the Effects of Cormorant Predation on Fish Populations

When the appropriate data have been obtained, researchers construct theoretical models to investigate biological interactions and the effect of cormorant predation on fish populations. Modeling approaches can range from simple, single-species models to highly complex exercises that attempt to account for the full complexity of a food web. Typically, most fisheries models focus on the target species of the fishery. In addition to food web interactions, a model has to account for the compensatory responses of fish populations that buffer the effects of particular kinds of predation. For instance, increased predation on adult fish may lead to increased growth rates and higher survival rates of young fish. Likewise, in the absence of predation, many fish would die anyway from competition for food or other population-density-related factors. When the levels of mortality occurring in a population would have occurred even in the absence of a particular mortality factor, mortality is described as compensatory. In contrast, when mortality is not buffered by compensatory processes and individuals are not replaced, mortality is described as additive. Physical factors like weather, the oxygen content of the water, and pollution often determine the amount of additive mortality in fish populations. In some situations, predation can also constitute additive mortality and remove fish that would otherwise be present. For example, this can occur if predators remove older fish directly from the fishery or, conversely, if they remove juvenile fish that have survived through the first year of life but have not yet recruited to the fishery. Predators can have additive effects when they consume secondary prey species too, but the predators' abundance is independent of the secondary prey's density.[12]

Relative to cormorant predation, these complex dynamics are sometimes boiled down to one $64,000 question: does the effect of cormorant predation on a fish population constitute compensatory or additive mortality? In reality, the effect of cormorant predation falls along a gradient between compensatory and additive processes, and shifts along that line in relation to factors influencing fish survival at any given time. Therefore, a rigorous scientific assessment of cormorant impacts requires identifying both where cormorant predation falls on the scale of additive versus compensatory mortality, and the level of mortality from cormorants that exceeds fishery goals (that is, diminishes the number of fish available to sustain desired levels of human fishing).

Calculating Changes in Fish Populations, Fish Harvests, and Economic Loss

The next step in the process is to calculate changes in fish populations resulting from predation by cormorants. Effects on the fishery can then be determined by calculating changes in catch or harvest that have resulted from the observed changes in the fish population. Of key importance is the calculation of economic losses attributable to these changes. Data on declines in catch or harvest resulting from historical increases in cormorant populations are also important indicators of impacts; so too are data on changes in catch or harvest that can be tied to control measures used to reduce cormorant populations. Finally, natural variability in all of the above processes should be considered when evaluating impacts.

Some Conclusions about Impacts to Open-Water Fisheries

The outline above provides a glimpse into the complexity of cormorant-fishery interactions and illustrates why assessing impacts is so difficult. Not surprisingly, few studies worldwide have been able to meet the substantial data requirements needed to determine the influence of cormorant predation on prey dynamics and fishery yields.[13] Nevertheless, the vast majority of cormorant predation studies report that the birds generally have inconsequential effects on fish populations, a conclusion frequently based just on diet assessment and information on factors affecting fish. For instance, an extensive review of cormorant food habit studies published by the USFWS in the late 1990s included twenty-five major studies that each reported results based on a minimum of thirty diet samples. As noted in chapter 8, the review indicated that

From a fish's view

important commercial and sport-fish species made up a very small proportion of the cormorant diet, and that cormorants have only a minor effect on fish populations compared to human harvesting and other mortality factors.[14]

A look at twenty-three additional studies published between 1995 and 2013 provides a similar conclusion (table 12.1). Eighteen studies directly evaluated the cormorant diet, while five estimated impacts by using modeling exercises, correlations between fishery trends and cormorant abundance, or information on cormorant foraging locations. Fourteen indicated that cormorants had only a minor or inconsequential effect on sport fish, sometimes even when consuming large numbers of desirable fish, or that sport or commercial fish species were an insignificant portion of the cormorant diet. While most of those studies did not meet the substantial data requirements needed to determine definitively how cormorants affected fish populations, they typically obtained data on key factors that supported their conclusions, such as observations of limits to predation across different times or places, the cormorant's position in a food web, or compensatory responses buffering predation (determined through modeling exercises). Moreover, of the four studies that that did not directly determine cormorant impacts, three documented prey switching by cormorants, which likely buffered predation impacts, and in one, fishery trends improved in spite of large numbers of cormorants feeding.

In general, support for a conclusion that cormorants are not causing negative impacts may be easier to obtain than for the opposite conclusion. This is

Table 12.1 Studies examining cormorant diet or impacts, 1995–2013

Study	Location	Assessment focus	Economically important fish consumed?	Consequential effect?
Kirsch 1995	Upper Mississippi River	Diet sampling	No	No
Madenjian and Gabrey 1995	Western Lake Erie	Modeling	Not directly assessed	Not directly addressed, but very minor relative to walleye predation
Milton et al. 1995	Nova Scotia	Diet sampling	Mostly no	Mostly no; not clear for Atlantic cod
Glahm et al. 1998	Lake Beulah, MS, and Lake Eufala, AL	Diet sampling	Mostly no	Mostly no, with possible exception of bluegill
Rail and Chapdelaine 1998	Gulf and estuary, St. Lawrence River	Diet sampling	No	No
Bur et al. 1999	Western Lake Erie	Diet sampling	No	No
Simmonds et al. 2000	Oklahoma reservoirs	Modeling	Yes, but not directly assessed	No

(continued)

Table 12.1 (*continued*)

Study	Location	Assessment focus	Economically important fish consumed?	Consequential effect?
Burnett et al. 2002	Eastern basin, Lake Ontario	Diet sampling	Yes	Possibly
Lantry et al. 2002	Eastern basin, Lake Ontario	Diet sampling	Yes	Yes
VanDeValk et al. 2002	Oneida Lake, NY	Diet sampling	Yes	Yes
Fenech et al. 2004	Lake Chicot, AR	Diet sampling	No	No
Rudstam et al. 2004	Oneida Lake, NY	Diet sampling	Yes	Yes
Seefelt and Gillingham 2006	Beaver Archipelago, Lake Michigan	Overlap between cormorant foraging locations and small-mouth bass habitat	No	No
Meadows 2007	Green Bay, WI	Diet sampling	Yes	Not reported, but fishery trends improved during study period

Study	Location	Method	Economically important fish	
Fielder 2008	Les Cheneaux Islands, Lake Huron	Fishery trends and cormorant abundance	Yes (Diana et al. 1997)	Yes
Heinrich 2008	Lake of the Woods, MN	Fishery trends and cormorant abundance	Not directly assessed	No
Seefelt and Gillingham 2008	Beaver Archipelago, Lake Michigan	Diet sampling	No	Not reported
Dalton et al. 2009	Bride Lake, CT	Diet sampling	Yes	No
Eisenhower and Parrish 2009	Lake Champlain	Diet sampling and foraging locations	Yes	No
Johnson et al. 2010	Eastern basin, Lake Ontario	Diet sampling	Yes	Unlikely; cormorant diet shifted to goby
Doucette et al. 2011	Saskatchewan lakes	Diet sampling	Some yellow perch	No
DeBruyne et al. 2012	Lake Champlain	Diet sampling	Some yellow perch	Undetermined; cormorant diet shifted largely to alewife
DeBruyne et al. 2013	Oneida Lake, NY	Diet sampling	Some	Unlikely; cormorant diet shifted to shad

Note: None of the studies in this table were included in the review by Trapp et al. 1999. "Economically important fish" refers to the question whether the fish eaten by cormorants at that location were significant for sportfishing or other economic activities.

because data on fishery trends, cormorant foraging locations, and cormorant niches may be relatively easy to obtain and may provide fairly conclusive evidence that cormorants are not causing impacts. But even in cases in which fisheries are declining and numerous circumstantial data point to cormorant predation as a potential cause, correlation does not equal causation. Instead, further analyses are required to reveal whether the cormorant is the real cause of declines. To date, published studies from just three areas (table 12.1) have concluded that fishery declines are due to cormorants. To varying extents, these studies incorporated the kinds of data and analyses necessary to describe how impacts occur. For the lessons learned and their influence on management actions, they are generally regarded as the most significant studies undertaken. By contrast, these studies also highlight more problematic situations in which management was based on far less scientific information (discussed in the next chapter). Nevertheless, this does not mean that the conclusions of even the most comprehensive studies are beyond question. Thus, these studies provide illuminating cases in which to examine the intricacies of cormorant-fishery assessment.

Three Case Studies
Oneida Lake, New York

Researchers from the Cornell Biological Field Station have studied interactions between cormorants, anglers, walleye, and yellow perch on Oneida Lake since 1988. Work by the fisheries scientist Lars Rudstam and his colleagues published in 2004 elucidates interactions between double-crested cormorants and their prey with a level of detail that to date no other study has been able to duplicate. The evaluation required long-term, continuous data on several age classes of fish, and because monitoring has been ongoing since 1957, many unique data were available. Numbers of adult walleye and yellow perch were estimated through the marking and recapturing of fish from the 1960s to 2001. Fish were aged from scales, and populations were estimated by age classes. Regurgitations, stomach contents, and pellets collected between 1995 and 2000 provided data on the species, size, and age of fish consumed by cormorants. Daily fish consumption was estimated from published data on cormorant energetics, and the proportion of the cormorant diet made up of each age group of perch and walleye was determined. The results showed pronounced negative differences in predicted versus observed recruitment for both perch and walleye after 1990, years when large numbers of cormorants were present on the lake. The difference between the predicted and observed recruitment was roughly comparable to the number of subadult fish (juvenile fish that survived their first year but were not of a size and age to be harvested) that cormorants consumed in the 1990s. The authors concluded that impacts occurred through predation on fish that were beyond the size range at which compensatory mechanisms would have a buffering effect, and thus that predation contributed to additive mortality.

The study included numerous factors that are challenging to estimate, and the authors noted uncertainty associated with estimates of both fish abundance and cormorant consumption. Nevertheless, they argued that despite these uncertainties, the data they obtained supported their hypothesis. In addition, several system-wide changes occurred in Oneida Lake in the 1990s that could also have affected fish populations, including an invasion by zebra mussels, increased water clarity, declines in nutrient inputs and phosphorus concentrations, and a decline in gizzard shad, an alternate prey for both walleye and cormorants. Considered in concert, these changes suggest that Oneida Lake may no longer be able to support the abundance of walleye and perch that it had in past decades. But when this study was published, associated effects resulting from these changes had not been documented or were not thought to explain the increased mortality of subadult fish.

Management to reduce cormorant predation has been ongoing since 1998. The Cornell research group has continued to evaluate the fishery, and two Cornell doctoral students, Jeremy Coleman and Robin DeBruyne, focused much of their research on predator-prey dynamics between cormorants, walleye, and yellow perch. These efforts indicate that, to date, walleye have responded positively during the period of cormorant management. But the magnitude of increase in subadult fish survival was larger than expected, based on the degree of decline in cormorant numbers and associated feeding days. Yellow perch, however, have not increased to the same degree. Although the pattern of decline ended, perch have experienced lower than expected recruitment, possibly from heavier predation from the increasing walleye population. Other changes in the lake may also explain or contribute to the observed fishery responses. For instance, gizzard shad became very abundant in 2001, and from 2001 to 2009 the species had the highest relative importance in the cormorant diet of any prey species consumed. In years when these fish are abundant, cormorants exhibit prey-switching behavior to consume them, and predation on walleye and yellow perch is buffered. Additionally, the stocking of walleye fry may have contributed to walleye recovery.[15]

Of further note is that after the invasion of zebra mussels, water clarity increased and may have made fish less vulnerable to trawls. Coleman's work suggested that part of the apparent walleye recovery, and possibly the recruitment dynamics of yellow perch after cormorant control began, could be an effect of underpredicting the number of fish projected to recruit because of low catches of year-old fish in summer trawls. Additionally, walleye may have benefited from the increase in gizzard shad. Overall, Coleman's discussion of trophic interactions highlights the complexity in evaluating not only impacts but also the effects of management.

The Oneida Lake studies made a strong case for cormorant predation as a major factor in the decline of walleye and yellow perch during the 1990s. Important parameters were measured directly, and most assessment methods were not associated with significant biases. As a result, these studies were able to provide robust empirical support for their conclusions. Moreover, Oneida Lake is a relatively small inland body of water, has an intensive stocking program for walleye fry, and is dominated by sport rather than rough fish, features that magnify impacts. But the possible problem with trawl data may limit the ability to determine the amount of fishery recovery attributable to cormorant management at this location. Similarly, the more recent changes in the lake's composition of fish species have led to the use of alternative prey by both cormorants and walleye, a potentially important factor in the fishery's recovery. And because the system has changed since the study was done, previous results cannot be extrapolated to present-day Oneida Lake, highlighting the importance of ongoing and site-specific monitoring. Finally, these observations suggest that cormorant control may not be necessary when alternative prey species are available.

The Eastern Basin of Lake Ontario, New York Waters

The New York State Department of Environmental Conservation has conducted annual assessments of the warm-water fish community in the eastern basin of Lake Ontario since the mid-1970s, obtaining data on the age composition, year-class strength, recruitment, survival rates, and growth rates of particular fish species. Annual efforts to quantify the cormorant diet began in

1992, and in 1998 an intense effort was made to compile the available information and evaluate cormorant impacts on smallmouth bass and yellow perch. The multidimensional work resulted in numerous publications during the late 1990s and early 2000s. Observations indicated that cormorants were targeting subadult fish of both species, suggesting that predation was resulting in a recruitment bottleneck. Researchers concluded that cormorant predation may have played a key role in maintaining low perch numbers, and appeared substantial enough to have caused the observed declines in the smallmouth bass population.[16]

Studies in the eastern basin provided figures for several parameters that are challenging to estimate, but some methods may have introduced important biases. Assessment of the cormorant diet and the age of fish consumed relied largely on pellet analysis, and estimates of the number of fish consumed incorporated the assumption that a pellet represented daily food intake. Additionally, estimates for bass and perch population sizes were not based on direct measurements. Instead, estimates from other areas and time periods were expanded and extrapolated, a common approach in natural resource management, but the practice introduces another level of uncertainty and potential error. Despite these issues, researchers believed their estimates projected realistic measures of fish population size and provided a conservative context for cormorant predation. Dramatic ecosystem changes similar to those reported for Oneida Lake also occurred in Lake Ontario as fish declines were observed, but researchers asserted that predation provided a more logical explanation for increased mortality than changes in the lake's food web and nutrient content. In addition, numbers of walleye increased during this time period, and this potential influence on decreased perch abundance could not be ruled out. However, the absence of smallmouth bass in a large sample of walleye stomachs collected over time led researchers to assume that walleye were not an important source of bass mortality. Another important factor unrelated to birds or ecosystem dynamics is the substantial underreporting of commercial harvests—discovered by an undercover operation conducted in 1997 by the USFWS and the New York State Department of Environmental Conservation—which may have contributed to increasing mortality rates for yellow perch in the 1990s.[17]

These studies were influential in the decision to initiate cormorant management in the region, which has been ongoing since 1999. But because of biases and the potential error associated with both pellet analyses and the extrapolated population sizes of fish, and also because of the influence of underreported commercial harvests, it is unclear how accurately researchers estimated cormorant consumption rates and the proportions of fish being removed by cormorants. For instance, the researchers doing the yellow perch work acknowledged that the majority of pellets they measured were eroded to some degree, and that this could bias results in two ways: "Erosion would cause the yellow perch length estimates to be shorter, and therefore younger than they actually were. This could lead to conservative loss estimates when the ages of yellow perch consumed by cormorants were related to estimates of standing stock. Otoliths from small perch also may have frequently been lost to erosion, which could bias consumption estimates to older ages. This would exaggerate the impact from cormorant predation."[18]

Moreover, earlier results are no longer applicable to eastern basin dynamics because the system has changed. And as in Oneida Lake, recent changes complicate the evaluation of cormorant management; the effectiveness of cormorant control in restoring fish populations has not been disentangled from other potential factors. One involves the invasion of round goby, a small soft-bodied fish native to Eurasia. By 2005, goby had become the primary prey of cormorants in the eastern basin, a dietary shift that went along with substantial declines in the amount of yellow perch and smallmouth bass consumed by cormorants. Population goals for cormorants at Little Galloo were reached by 2006; that achievement, combined with the dietary shift to round goby, suggests that predation pressure on smallmouth bass has been substantially reduced. Furthermore, round goby became an important prey item for smallmouth bass too, which has contributed to improved bass growth rates and condition. The smallmouth bass catch per unit effort, an indirect measure of abundance, increased in 2005 and has been stable ever since, indicating some improvement in the fishery. This may have resulted from both reduced predation and increased catchability, the latter arising from improved growth and conditions producing greater average body lengths at younger ages. But despite these changes, in 2012 there was still no evidence of any strong year-class recruiting into the older age groups, indicating that the smallmouth bass population had not fully recovered. Finally, the ability to evaluate the effectiveness of cormorant management for yellow perch recovery has become even more difficult, since excessive quantities of invasive mussels have ended the bottom-trawling survey necessary to assess the perch population in the basin.[19]

Les Cheneaux Islands, Michigan

Work to evaluate the role of cormorant predation in the declining yellow perch population of the Les Cheneaux Islands region began in 1995, when a one-year field study was undertaken by the fisheries scientist James Diana, from the University of Michigan, and biologists from the Michigan Department of Natural Resources. To date, this has been the only study on the Great Lakes to simultaneously measure cormorant population size and diet, fish population size and mortality, and sport catches. The cormorant diet and the age of fish consumed were determined by examining stomach contents of cormorants shot shortly after they were observed feeding. Daily food consumption was estimated based on published information and the caloric density of fish consumed. The population size of yellow perch was estimated by marking and recapturing fish, and information on perch size and age structure was compiled and based on fish caught in nets, in trawls, and by anglers, and on tagged fish. Annual perch mortality was estimated through three analyses. To estimate angler-caused mortality, a creel survey was used, and the overall fishing effort was estimated from counts of fishing boats throughout the season. Results were summarized in reports and technical bulletins in the late 1990s and officially published in 2006.[20]

Over the season, alewife was by far the most important cormorant prey item, while yellow perch was important only when spawning, during April and May. Most perch consumed were subadults, and predation by cormorants was estimated to account for 1 percent of mortality experienced by legal-sized fish. Similarly, anglers accounted for 2.4 percent of mortality. The total annual mortality rate for the yellow perch population was estimated at 45 percent, indicating that other mortality sources remove roughly 40 percent of the legal-sized fish. The authors concluded that while cormorants were highly visible predators, they did not constitute a major source of mortality in 1995. Further, they suggested that cormorant predation on perch predators and competitors could benefit perch. But the authors cautioned that there could be important year-to-year variations in the timing of perch spawning, in the cormorant diet, and in sources of mortality for perch. They did not determine whether cormorant predation on fish smaller than legal size constituted additive or compensatory mortality, but at current predation rates, they speculated that even if cormorant predation represented additive mortality, the effect on the future abundance of large fish would not be substantial. Although no estimates of young-of-year perch were obtained, based on the size of other year classes recorded for the area, the authors argued that cormorants would likely not have substantial impacts on recruitment. Finally, the authors put perch population declines into a regional context, noting that depressed perch populations had been observed in many other areas of the Great Lakes.

One possible bias of this study is that the large open-water system of the study area may limit the usefulness of the mark-and-recapture method used to estimate perch population size. Additionally, reliance on tag-return information from anglers could introduce bias if an unknown proportion of tags went unreported. But the authors asserted that all assumptions implicit for a population estimate based on mark and recapture were met. Overall, the study met the data requirements necessary to provide strong support for the conclusion that cormorants were not a major factor contributing to yellow perch declines in 1995.

But the story there was just beginning. Five years after Diana's study, the perch fishery essentially collapsed. To explore potential causes, preliminary work was undertaken by Dave Fielder, a fisheries biologist with the Michigan Department of Natural Resources. This work, published in 2004, examined data on trends in yellow perch and cormorant abundance, and the annual mortality and recruitment of perch. Limited age data indicated that fish continued to be recruited during the fishery's decline, and that the mean age of fish in the population had decreased. Additionally, despite declines in fishing pressure, mortality rates remained high. Meanwhile, numbers of cormorants remained abundant during this period of decline, but growth had slowed considerably since 1995, suggesting population size was leveling off. Comparing his results to Diana's, Fielder noted that both yellow perch and alewife had been more abundant in 1995 than they were during 1997–2002. Declines in alewife in the late 1990s raised the question of whether cormorant predation on yellow perch may have been greater during the later period in order to compensate for the diminished numbers of alewife. If so, Fielder suggested that "such a predation scenario would account for the continued high total annual mortality rate and the decline in mean age" of perch. His interpretation of the data he analyzed was that "the collapse of the fishery and range contraction of perch were caused at least in part by the predatory effects of cormorants." While noting that additional research was needed to better quantify the role of cormorant predation in this decline, he stated that "in the absence of other explanations it is compelling to conclude that cormorant predation is at least part of the explanation." Notably, Fielder's analysis did not evaluate the current cormorant diet but instead built on Diana's 1995 study to support the assumption that cormorants were consuming yellow perch. Although the 1995 study indicated that perch was a relatively insignificant portion of the cormorant diet compared to alewife, the possibility that the relation between cormorants and perch could have changed by the later period in the way that Fielder described significantly influenced the decision to launch a cormorant management program in 2004.

In 2008, Fielder published an expanded analysis building on his earlier work. It documented that strong year classes of perch had been produced even during the decline, but the fish disappeared quickly. This suggested that predation on young fish could be leading to the same kind of recruitment bottleneck as that observed in the New York studies. Although mortality rates in Fielder's analysis mostly reflected fish two years old or older, whereas the 1995 diet work demonstrated that cormorants fed almost entirely on juvenile fish, Fielder speculated that cormorants may have consumed perch of a greater size and age range as fish became scarce. He attributed the yellow perch collapse to excessive mortality rather than to declining recruitment, concluded that cormorant abundance was the most influential force on the perch fishery and population during the time period evaluated, and hypothesized that cormorant control would lead to improvements in the fishery.

Fielder's conclusions provided continued support for management, which, by the time the 2008 paper was published, had already reduced cormorant numbers in the Les Cheneaux Islands by 70 percent.[21] In light of significant disagreements with Fielder's conclusions, and the management for which they were the basis, Jim Diana published a commentary on Fielder's work in the *Journal of Great Lakes Research* in 2010. Diana began by identifying numerous system-wide changes in Lake Huron since 1995 that could have caused a decline in perch, including large declines in alewife, Pacific salmon, and the small crustacean *Diporeia* (an important prey item for many forage fish); invasion by round goby; changes in exotic mussel populations; and changes in temperature conditions and water levels due to an evolving climate. He then went on to identify "variable population trends, uneven data analysis, and limited data collection" as major flaws in Fielder's analysis, arguing that if cormorants had caused the perch collapse, then young fish, which cormorants eat, should have declined first, rather than older fish, which they typically ignore. Further, if the cormorant diet had expanded to include larger perch, the birds would likely have continued to eat smaller ones too, and the collapse would be a result of both increased adult mortality and reduced recruitment. Because Fielder did not evaluate the cormorant diet or obtain direct evidence for an increase in the size or number of perch eaten, Diana argued that Fielder had not provided a logical explanation of how cormorants increased the mortality of adult perch. He also described similar trends in perch harvests from regions in Michigan where the number of cormorants ranged from none to several thousand. Based on these data, he wrote, "it appears that factors other than cormorants are influencing perch harvests and that these factors are linked to region-wide characteristics throughout the Great Lakes." To conclude, Diana provided a philosophical discussion and addressed issues that he believed precluded an ethical justification for cormorant control on the Great Lakes. He stated that as an ecologist who studies fish populations, he had strong concerns about killing cormorants solely to improve fisheries and considered such management to be "philosophically wrong, except in locations with very clear indications of overwhelming damage from the birds."

Fielder's response, published in the same volume of the journal, asserted that his work built on Diana's, which had already established that cormorants consume yellow perch. He defended his methods and conclusions, tossed some questions back to Diana, and suggested that the perch mortality rate in the 1995 study may have been underestimated. He referred to data in his 2008 paper showing that the mean age of fish in the Les Cheneaux Islands had steadily increased from the mid-1980s to the mid-1990s, and then abruptly dropped. He argued that cormorants had likely been consuming young perch first and then moved on to older perch after young perch became scarce. Additionally, he pointed to a series of forthcoming publications demonstrating that cormorant control had led to a rebound in the yellow perch fishery and was "widely heralded as a management success." In regards to the ethical issues Diana raised, Fielder characterized this portion of Diana's commentary as a "manifesto on his personal philosophy against cormorant management in the Great Lakes," and stated that as a researcher he preferred to leave such "policy matters . . . to decision makers in management agencies."[22]

Fielder's evaluation of the yellow perch recovery was published in the next issue of the journal. His results indicated that as cormorant numbers declined, the abundance and mean age of perch increased, total mortality and growth rates decreased, and angler catches and angler harvests improved. Perch recruitment also increased but Fielder noted that the true level of recruitment may not have been that much greater than in previous years; rather, young fish released from cormorant predation may have experienced better survival, giving the appearance of better reproductive success. Based on his results, Fielder concluded that cormorant control activities met their objective. He suggested a sustainable cormorant population size for the region, from a fisheries allocation perspective, might be well below 1,000 birds or 500 nests, requiring at least a 90 percent reduction in numbers from the population size that existed when control was initiated.[23]

Overall, Fielder's work provided a careful examination of the yellow perch population, fishery trends, and some factors influencing them. His analyses present a convincing explanation of how cormorant predation may be affecting the yellow perch population. Additionally, the timing of the yellow perch recovery coincided with the initiation of cormorant control in the region—which culled adults and oiled eggs—and helps strengthen Fielder's conclusion. But as Diana pointed out, Fielder did not provide direct evidence to support the assertion that during the later period of perch decline, the cormorant diet expanded to include larger and more perch. Without this kind of empirical data—an arguably key component of impact assessment—a definitive answer remains elusive.

While the yellow perch recovery coincided with cormorant management, it is also possible that other changes in the lake and in the cormorant diet contributed to the recovery, or can at least partly explain it. For instance, improvements in yellow perch abundance and the sport fishery began to be observed as early as 2004 and 2005, and the mean age of the fish began to increase as well. But because previously obtained data suggested that cormorants fed primarily on young perch when management was initiated, relieving cormorant predation on that age group would not be expected to affect anglers' harvests for at least two years. Therefore, the early changes observed with the initiation of cormorant control suggest that perch may have been experiencing some level of recovery unrelated to cormorant reductions. A concurrent general improvement in the perch fishery in Lakes Michigan and Huron in areas where no cormorant management was undertaken supports this idea. Furthermore, large improvements in yellow perch reproductive success occurred simultaneously in areas of Lake Huron where cormorants were not managed; that recovery was linked with some of the widespread changes the lake had recently undergone. Finally, lessons learned from the detailed monitoring at Oneida Lake and the eastern basin of Lake Ontario indicate how other factors may influence fishery trends concurrent with cormorant management. As a result, the evaluation of fishery trends in relation to cormorant abundance may not by itself be enough to demonstrate the effectiveness of cormorant management in meeting fishery objectives, unless the influence of other factors can be ruled out.[24]

The focus of this chapter was to examine the science illuminating cormorant-fishery interactions, and to provide some idea of the complexity involved in impact assessments. Several take home messages emerge. First, assessing cormorant impacts requires a variety of data on both cormorant and fish populations. Second, the effects of cormorant predation at any given time and place may be a moving target, since systems are dynamic and cormorants are adaptive; these observations highlight the need for ongoing monitoring to understand the role of cormorants in fishery conditions. Third, impacts result from site- and time-specific conditions, and so results from one location or time period should not be extrapolated to another. Fourth, even the most thorough studies documenting cormorant impacts contain some level of uncertainty, and the strength of their conclusions must be weighed against the limitations inherent in their methods and study design. Fifth, determining the effectiveness of cormorant management, even if careful monitoring of the required factors is undertaken, may be confounded by simultaneously occurring changes in aquatic systems. These observations lead to the next key question in the cormorant's story: in light of so much uncertainty, how has cormorant management to reduce impacts on fisheries proceeded, and have management actions been reasonable and appropriate?

13

Adaptive Management

A Process Gone Awry

Sometimes there is no evidence at all. Two hundred years ago, ed-
ucated people imagined that the greatest contribution of science
would be to free the world from superstition and humbug. It has
not happened. Ancient beliefs in demons and magic still sweep
across the modern landscape, but they are now dressed in the language and symbols of science.

—Robert L. Park, *Voodoo Science* (2000)

THE AMBIGUITY SURROUNDING cormorant impacts on fisheries and the
management actions to reduce them is in no way exceptional. Rather, the
management of ecological systems and natural resources is frequently beset
by significant uncertainty about the system being managed and how it will be
affected by management actions. An approach commonly taken in such situations, and one now widely invoked to justify a broad range of actions taken to
resolve cormorant-fishery conflicts, is adaptive management. This approach
focuses on actions that facilitate learning while doing. It involves a decision
process that uses findings from studies and experiments to direct the course
of future management activities, and is characterized by several key components: careful biological monitoring that recognizes the natural variability in

ecosystems; management actions designed to reduce uncertainties about ecological relationships that drive resource dynamics; repeated measurements of the effects of management actions; quantitative modeling to make predictions about alternative management options; and the refinement of management actions as more data become available. Adaptive management is not an end in itself, but a process that is essentially experimental in nature, geared toward increasing the knowledge of a system in a way that leads to more effective and informed decisions.[1]

In theory, adaptive management can be used with any sort of decision-making process that a management context permits. But in practice, the term has been inappropriately applied so frequently that several recent publications have become available to define adaptive management more specifically and identify important challenges and opportunities for its implementation. Of special note is a white paper of 2011 by the Center for Progressive Reform, a nonprofit organization comprising a network of US scholars focused on protecting health, safety, and the environment through analysis and commentary. The paper observed that adaptive management as practiced by natural resource agencies, in many cases, "has become at best uninformative and at worst a smokescreen for unbounded agency discretion." To ensure that the approach is not misused, the paper identified three prerequisites that must be met for the approach to be applied: the existence of information gaps that need to be addressed in order to achieve management goals; the possibility of narrowing those gaps over a "management-relevant time scale"; and the possibility of adjusting initial management actions as new data become available.[2]

The ability to develop an effective monitoring program is also paramount for the approach to be applicable. In this regard, Byron Williams, a scientist with the US Geological Survey, identified in 2011 the issue of "nonstationary resource dynamics" as an especially difficult challenge for adaptive management. The ability to evaluate resource trends in response to management actions rests largely on the assumption that both the resource system in question and the underlying environmental conditions are dynamically stable over the management time frame. But in many situations, environmental conditions are evolving, and the targets of study—the ecological processes that determine resource change—are evolving as well. Williams identified several ways to address this "nonstationarity." One includes tracking and modeling the environmental drivers of change, and then using trends in environmental conditions to account for the changes that resulted in resource patterns over time. Another involves looking for time frames over which resource dynamics are largely stationary. A third develops scenarios of directional change that are based on assumed patterns of directionality, with adaptive decision making accounting for uncertainty about which scenario is the most appropriate. At a minimum, the approach must consider directional trends in environmental conditions and resource dynamics, and make efforts to accommodate them.

Adaptive management takes on another level of complexity when it is applied to the control of nuisance wildlife, which typically involves the destruction of large numbers of animals in order to solve human-wildlife conflicts. To identify the value of the approach in situations of this nature, the scientist Bruce Warburton and the philosopher Bryan Norton published an important commentary in 2009. The authors describe the control of nuisance wildlife as a classic example of a "wicked" problem for wildlife managers, one that involves competing resource goals and human values, such as the protection of fisheries versus the protection of birds, and that has no single correct solution. For such problems, they identify the need to consider decisions "within a pluralistic framework of values that considers suffering and death of sentient creatures but considers other, competing, values as well." Because the management of nuisance wildlife often has a large component of uncertainty, especially when multiple species or density-independent factors influence the system, Warburton and Norton propose embedding control operations within an experimental, adaptive-management framework. Doing so is "the only ethically defensible action," since it leads to future management actions that are carried out with increased understanding, and thus "contains the seeds for a knowledge-based ethic for managing nuisance wildlife." Besides reducing uncertainty, the approach can be further justified because it "exposes poorly conceived projects to test, and stimulates discussion of social values." Finally, they emphasize that such management must make a "heavy investment" in "good" science and ensure that animals are not "killed wantonly." And therein lies the rub: such management requires complex monitoring in order to evaluate the effects of manipulating animal numbers, and makes large-scale experiments difficult and expensive to conduct. As a result, Warburton and Norton acknowledge that "the cheaper option of *just killing nuisance wildlife* will often be the favored choice of action" (emphasis added).

This discussion is particularly germane to the management of cormorants. The recent extermination of tens of thousands of cormorants and their eggs in the Great Lakes region has been presented in the context of "scientific experimentation" that will lead to knowledge-based solutions. But thus far the scale of the "experiment" makes it virtually indistinguishable from the solution long advocated by some segments of the public and certain wildlife professionals: massive culling and other efforts to reduce cormorant numbers. This poses the question, are current management practices truly adaptive programs that lead to knowledge-based solutions, or do they ultimately comprise programs of "just killing nuisance wildlife"?

Wildlife Acceptance Capacity

Contemporary wildlife management increasingly reflects the involvement of stakeholders and members of the public in the decision-making process, and attempts to integrate social issues with the biological dimensions inherent in human-wildlife conflicts. As a result, engaged citizens can significantly influence management objectives—and many of those involved in cormorant conflicts dislike cormorants and perceive them to be overabundant. And perhaps because cormorants are such striking birds, it does not take many of them to fuel an impression of large population size and convey a sense of there being "too many." But almost always, such a judgment is a matter of perception rather than reality. In a scientific framework, the concept of overabundance is directly related to the biological carrying capacity of the environment, which defines a maximum population size that can be sustained indefinitely without degrading the resources necessary to support individuals. For cormorants, there have been very few observations of populations reaching carrying capacity (see the discussion in chapter 3), and no studies have demonstrated that the species is overabundant relative to biological resources.[3]

But there are other frameworks in which the concept of overabundance has nothing to do with biological carrying capacity. In 2003, a paper prepared by the Wildlife Services biologists Jimmy Taylor and Brian Dorr stated that cormorant populations, "though operating under biological carrying capacity, have exceeded acceptance capacity with several wildlife stakeholder groups throughout Canada and the United States." "Wildlife acceptance capacity" is a concept for wildlife management that developed in the late 1980s. It prioritizes relationships between people and the environment, so populations are considered overabundant when they exceed human tolerance levels. Predictably, animals that come into conflict with humans typically have wildlife acceptance capacities that are far lower than what the biological carrying capacity of the environment could actually support. For cormorants, which have far fewer advocates than detractors, increased stakeholder involvement and the concept of wildlife acceptance capacity have especially important implications for how they are managed. This is especially relevant in light of the decidedly biological solution—population reduction—applied to a problem that has thus far eluded characterization as primarily social, ecological, or economic.[4]

Management at Oneida Lake, Lake Ontario's Eastern Basin, and the Les Cheneaux Islands

To evaluate the appropriateness of cormorant management, three main criteria are considered: the evidence indicating that management was needed in the first place, the prerequisites for adaptive management, and the degree to which programs have been conservative and balanced. The last is especially important because one reasonable expectation for programs employing experimental actions that are irreversible or destructive, such as culling and egg oiling, is that they be as conservative as possible—in other words, animals should not be "killed wantonly." To provide guidance on this issue, the Center for Progressive Reform recommends that such actions be undertaken incrementally, allowing an opportunity to learn from and correct mistakes before impacts become too severe. A related expectation is that programs be balanced, since they are attempting to address competing resource interests, in this case cormorants versus fish. Thus, to evaluate whether programs have been conservative and balanced, close attention is paid to the magnitude of population reductions and the time frame over which management proceeded. This exercise begins with the three areas that have thus far presented the most compelling evidence to demonstrate negative impacts on fisheries from cormorants. Although there were degrees of uncertainty about impacts in each location, there were enough scientific data from each area to justify some type of adaptive management action.[5]

At Oneida Lake, both breeding birds and fall migrants consume fish, but fall migrants pose the biggest problem. Initially, efforts concentrated on limiting reproduction through nest control and on dispersing migrants through intensive hazing. Between 2004 and 2009, efforts intensified: birds were harassed throughout the entire breeding and migration seasons, and all nests on the lake were targeted for destruction. In 2010, federal funding was lost for the program, which had been carried out by Wildlife Services since 1998. The program then shifted to only fall hazing and was conducted by the New York State Department of Environmental Conservation and volunteers.[6]

Overall, the program has been fairly conservative, in that actions to manage cormorants have not incorporated the destruction of birds, though they became progressively more restrictive, and eventually no cormorants were tolerated on the lake. From a fisheries allocation perspective, the level of management undertaken was appropriate during years when birds were focused on walleye and yellow perch. Additionally, efforts to assess the status of the fishery and the effectiveness of cormorant management have been comprehensive. Ongoing monitoring has included the collection of a limited number of cormorants each year to assess the cormorant diet, as well as analyses of trends in lake biology and of fish community changes.[7] But since 2001, gizzard shad has been abundant and, along with emerald shiner, has buffered the impacts of cormorant predation on yellow perch and walleye. This change in prey base and the associated prey switching by cormorants suggest that larger numbers of cormorants could be supported; thus, the need for the level of management initiated in 1998 has not been demonstrated in recent years. Nevertheless, management actions were not modified until 2010, and then only because federal funding was cut.

Under the more recent management scenario, only moderate increases of cormorants in summer months have been observed, and potential impacts from fall migrants were still prevented, indicating that this reduced level of management could be effective. But because of concern that current human harvest rates would not be sustainable if the cormorant population were to rebuild itself, fisheries scientists from the Cornell Biological Field Station recommended that efforts to limit cormorant numbers continue. Furthermore, the Oneida Lake Association continues to pursue restoration of the intensive federal program by Wildlife Services. This association has strong political influence on matters involving the lake's natural resources, particularly fisheries conservation, and has been a potent advocate for cormorant control. Along with fishermen, these stakeholders will continue to influence how management proceeds and potentially limit the extent to which management actions can be adapted despite changing fishery conditions.[8]

In the eastern basin of Lake Ontario, actions taken on Little Galloo Island were initially conservative, using only egg oiling and nest destruction to reduce cormorant numbers. The population objective, however, called for a decrease of 82 percent from the peak colony size in the late 1990s, which encompassed a reduction of thousands of birds. Additionally, after the establishment of the Public Resource Depredation Order, culling was added as a full-scale measure, because cormorant numbers, though declining, were not declining fast enough. Between 2005 and 2011, 10 percent of the breeding adults, on average, were shot annually, totaling close to 4,000 birds.[9]

Like the Oneida Lake program, efforts have been undertaken in the eastern basin to comprehensively assess the status of the fishery, the factors affecting it, and the cormorant diet. But some of the actions appear fairly liberal relative to the scientific studies that led to the initiation of management. For instance, as described in the previous chapter, there was significant uncertainty associated with specific estimates that could have influenced the degree to which cormorants were determined to impact fish. Additionally, the population objective was based on circumstantial data—the maximum numbers of cormorants present before declines in smallmouth bass began to be observed—rather than on a quantified link between numbers of cormorants and fish. Implementing such a large reduction in population size ignored this uncertainty and likely eliminated more birds than would have been necessary in order to see some improvement in the fishery. Furthermore, the move to lethal control occurred when the cormorant population was already significantly diminished from egg oiling.[10] And it was initiated as a full-scale measure in the same year that the cormorant diet began to be dominated by round goby, a change that should have indicated a need for less rather than more aggressive management.

Despite the significant ecosystem changes in the basin, and prey switching by cormorants to round goby, culling and egg oiling remain in practice, and the cormorant population objective has not been revised as of 2012. These observations suggest that management thus far has not been truly adaptive. As at Oneida Lake, significant pressure from stakeholders such as fishermen and Concerned Citizens for Cormorant Control will likely continue to be an important influence on management directions. The extremely low tolerance for cormorants shown by these stakeholders may make it difficult to increase the cormorant population target in the future, even if higher numbers of cormorants could be sustained without impacts on the fishery, as a diet dominated by round goby suggests would be the case.

Heading west to Michigan's Les Cheneaux Islands, intensive cormorant management was initiated in 2004 on three of the five islands where cormorants nested. The initial goal was to oil all ground nests and shoot 15 percent of breeding adults. By 2006, the management goal for culling adults had been increased to 50 percent, and all five islands were being managed. By 2007, 4,205 cormorants had been culled, and the population of nesting cormorants had been reduced by more than 70 percent. In 2008, a cormorant population objective of 500 pairs was announced, requiring a reduction of 90 percent from the precontrol population size. It is not apparent how this goal was selected, other than that it reflects the minimum colony size of 100 pairs that Michigan's environmental assessment specified must be allowed to remain at each colony site subjected to management. As of 2012, this goal had been reached; only 476 pairs remained.[11]

For several reasons, the approach taken in this region has been particularly problematic. First, as described in the previous chapter, a timely assessment of the cormorant diet and broader food-web interactions was not undertaken before management was initiated, and there is significant scientific debate about cormorant impacts in this area. Management actions and goals have nonetheless been extremely liberal, proceeding as though there were no doubt about the role of cormorants in fishery declines. The cormorant population objective was established to determine whether the perch fishery would be sustainable in the presence of very small numbers of cormorants, yet no peer-reviewed scientific work had demonstrated that such a massive and immediate reduction of breeding birds was necessary. To the contrary, the perch population began showing signs of recovery as early as 2004 and 2005, when between approximately 3,200 and 4,600 pairs of cormorants were present. In fact, in a 2008 newspaper story on cormorant control in the Les Cheneaux Islands, the perch population recovery between 2005 and 2006 was described as "explosive," an observation that suggests larger numbers of birds may have been sustainable.[12] But the only adaptation to management was to intensify control activities.

Second, sometime around 2002, raccoons were intentionally released onto Goose Island, which in 2003 supported more than one third of the region's breeding population. Shortly after the raccoons were introduced, birds began abandoning nests and breeding numbers declined. By 2006, the island had been abandoned, and the regional population had declined by 63 percent. Because the population was already declining in response to the raccoons, there was no need for more intense management. Nevertheless, culling continued on the island for as long as cormorants persisted, and control became more aggressive across the region.[13]

Third, the status of the yellow perch fishery is assessed annually, and some cormorants are collected for diet analysis, but efforts to evaluate the fishery response to cormorant management have relied largely on correlation analyses between improvements in the fishery and declines in bird numbers. But the system is dynamic and other factors may simultaneously influence perch abundance. For instance, by the mid-2000s round goby were established in the Les Cheneaux Islands area; because this species can buffer predation on perch, larger numbers of cormorants could potentially be supported without negative impacts to the fishery.

Finally, the 2011 environmental assessment for Michigan indicated that the population objective of 500 pairs will be maintained for five more years to determine whether the fishery will remain stable in the presence of this number of cormorants. The assessment gives no indication that larger numbers of cormorants would be tolerated if perch remained stable, but it stated clearly that "if fish population metrics indicate [continued] declines are probably attributable to cormorants, additional reductions may be considered." As in the New York areas, there is intense pressure to manage birds from fishermen, business owners, fisheries managers, politicians, and associations such as the Les Cheneaux Islands Sportsman's Club. These stakeholders will undoubtedly remain a formidable barrier to adapting management practices in the direction of larger cormorant numbers, even if they can be sustained. This observation, combined with the reasons just noted, indicates management practices here are not adaptive.

Management at Other Areas in the Great Lakes

If cormorant management in the three areas described above has not been truly adaptive, how then has it proceeded in the many other areas where far less science has been available to inform decisions? To answer this question, we must look at significant programs targeting cormorant-fishery concerns in other areas of the Great Lakes, specifically those that have been identified as taking an adaptive management approach. Common to all are a diverse array of stakeholders, including local businesses, resort owners, anglers, fisheries managers, fishing and wildlife clubs, and politicians with very low wildlife acceptance capacities for cormorants.

Michigan

Since 2004, management has been implemented at a minimum of thirty-two locations across Michigan. In addition to the Les Cheneaux Islands, three more management units encompassing multiple colonies have been targeted. The overall intensity of management has led the state to define a minimum state-wide cormorant population of 5,000 pairs, below which the population will not be allowed to drop. That population minimum represents an 83 percent reduction from when management was initiated, or a decline of about 50,000 breeding birds if it is reached. Below, actions in the three additional management units are summarized from the 2006 and 2011 Michigan environmental assessments and the required reports summarizing annual control activities submitted by Wildlife Services.

In Thunder Bay, Lake Huron, cormorants breed at four colony sites, and 3,994 breeding pairs were present before the initiation of control in 2006. Concerns revolved around declines of young whitefish in trawl catches; declines in brown trout, a nonnative species stocked into Great Lakes waters by the Michigan Department of Natural Resources; and declines in catch rates for all fish species vulnerable to the trawl. Causes for these declines were unclear, but important food-web changes associated with zebra and quagga mussel invasions in the 1990s were thought to have played a role, along with poor prey resources and declines in alewife. Additionally, the overall niche for brown trout in the region was believed to be tenuous. While highly unlikely that cormorants alone would cause declines in all fish species simultaneously, the Department of Natural Resources speculated that large numbers of cormorants could lead to adverse community-level effects.

In 2005, based on the number of cormorants assumed to be present, the agencies estimated that cormorants consumed more than one million pounds of fish, far exceeding the bay's estimated standing crop of fish vulnerable to the trawl. That exercise involved many untested assumptions and likely overestimated cormorant numbers by assuming that there was one nonbreeding bird for every pair present. Nevertheless, it indicated the potential impact that cormorants could have on fishery resources. Although the agencies lacked a careful evaluation of the cormorant diet that linked predation with declines in any fish species, concern that cormorants could diminish resources in one way or another was justification enough to establish a regional population objective of 450 pairs and to initiate culling at several locations. This objective required a reduction of about 90 percent of breeding birds, and was based on the number of birds present before declines in fish occurred, which also corresponded to a time before major changes in the food web were documented. Goals for the fishery included improved survival of stocked species and lake whitefish, along with reduced predation on forage fish. The media later characterized the cormorant population objective as "the magic number," and Jim Johnson, a Department of Natural Resources fisheries biologist, further justified the aggressive control measures by telling the media that cormorant numbers in the region were "anything but natural"; instead, they were "invasive numbers." He also noted that the response observed in the Les Cheneaux Islands was being used as a "template" for Thunder Bay. As of 2012, the population there had been reduced by 81 percent.[14]

In Big Bay de Noc and Little Bays de Noc, Lake Michigan, cormorants breed on four islands, and 9,854 pairs were present when control was initiated in 2006. Concern centered on perceived impacts on the fish community in general. The Department of Natural Resources undertook a preliminary analysis to estimate the potential impact of cormorant predation based on numbers of cormorants present in 2005. Estimates for standing fish biomass and total annual biomass that equated to total annual fish production relied on data from the North Channel of Lake Huron. Results indicated that cormorants were consuming 100 percent of the total annual production of the fish community and were exerting an unsustainable demand of more than four times the amount of annual production. Despite no actual data on the size of fish populations, no cormorant diet studies having been conducted in the region, and reliance on several untested assumptions, the exercise was sufficient to initiate cormorant management. Additionally, the agencies noted that cormorant consumption was increasingly being viewed as a matter of resource allocation, and that they were interested in making more fish resources available to both desired game fish and local fisheries. Goals included improving the walleye and yellow perch fisheries and reducing foraging pressure on the prey base. The objective for cormorants was to reduce the number of breeding birds by 50 percent a year until an improvement in fisheries was observed, and was strongly supported by local residents, fishermen, and wildlife and fishing clubs. While nearly 8,000 birds were culled between 2006 and 2011, intensive efforts to reach the 50 percent reduction goal were not undertaken until 2012 because the Department of Natural Resources wanted Wildlife Services to focus its limited resources on other areas in the state considered more urgent. As of 2012, the population had been reduced by 37 percent.

In the Beaver Island Archipelago in Lake Michigan, cormorants breed at five principal sites. The population totaled 11,400 pairs when control was initiated in 2007, at which time it was the largest population in the state. The Beaver Island Wildlife Club has been a particularly strong advocate for cormorant control and has thus far constituted a potent public influence on the need for cormorant management there. Fishery concerns developed primarily over potential impacts on the declining smallmouth bass fishery, and the agencies reiterated the need to view cormorant consumption from a perspective of resource allocation. Circumstantial evidence implicating cormorants as the cause of the smallmouth bass decline was presented in a master's thesis undertaken between 1999 and 2002 by Michael Seider at the University of Georgia, who evaluated the population dynamics of smallmouth bass. But his study did not measure the cormorant diet or examine cormorant habitat use. Concurrently, doctoral work undertaken by Nancy Seefelt at Central Michigan University evaluated cormorant population size, diet, and foraging behavior in the archipelago. After examining 150 cormorant stomachs, 978 regurgitated samples, and an unspecified number of pellets, Seefelt found only one smallmouth bass, indicating that cormorants almost never consumed this species. She also demonstrated through radio telemetry and observational study that birds from Beaver Island colonies spent relatively little time foraging in areas historically important to smallmouth bass. Her work indicated that the potential for cormorants to affect bass was low because of spatial separation. A final portion of Seefelt's thesis modeled the relationship between declining bass numbers and cormorants; it suggested that cormorant predation alone was unlikely to have caused the observed declines, nor was it the leading factor limiting bass population size in the region.[15]

However, the Michigan Department of Natural Resources questioned whether assumptions in the model accurately represented what could happen if the bass population started to recover, although forecast models were included in the thesis to address that issue. The agencies noted that even an extremely low occurrence of smallmouth bass in the cormorant diet could be detrimental to recovery, and suggested that if just 12 percent of cormorants each ate a single bass over the summer, the consumption would equal the estimated number of bass present between 1999 and 2002 and would thus eliminate the population. In 2007, culling and egg oiling were initiated at several locations, and through reductions of 50 percent annually, a population objective of 3,000 pairs is currently targeted. Fishery goals include restoration of the smallmouth bass fishery and reduction of the overall foraging demand on the prey base. As of 2012, the cormorant population had been reduced by 58 percent.

In the 2011 environmental assessment for the state, few new data of any substance were presented to support the continued need for cormorant management in these regions. Conversely, the stomach contents of cormorants shot in Thunder Bay and Bays de Noc indicated that round goby comprised 84–91 percent of the prey items consumed by cormorants in those areas. Similarly, the wildlife scientist Nancy Seefelt, who continues to study cormorants in the Beaver Island archipelago, provided data to the agencies (later published) indicating that the cormorant diet in that area likewise largely consisted of round goby. In her public comments on the assessment, Seefelt suggested that the lower nutritional value of round goby would likely lead to natural declines in the cormorant population. But neither her diet data nor that from the Bays de Noc or Thunder Bay altered management objectives for any of these areas. Management continued for the reasons already described, and also because there was the potential for the cormorant diet to become more diverse in the future as fish populations recovered. To assess management's effectiveness in each of these areas, monitoring focuses on comparisons between cormorant population reductions and fishery data obtained through surveys (for example, creel, trawl, and gill-net surveys). Additionally, limited numbers of cormorants are collected in some years to assess the cormorant diet.[16]

Wisconsin

In Wisconsin, two management units encompassing multiple colonies in Green Bay have been the focus of large-scale activities primarily to benefit fisheries. Commercial and sports fishermen have exerted strong pressure for the management of cormorants in the area, and after the Public Resource Depredation Order was established, complaints from those groups motivated the Wisconsin state legislature to establish Wisconsin Act 287, requiring the Department of Natural Resources to administer a cormorant management plan.[17] Management in these units is summarized from the 2009 Wisconsin environmental assessment and the 2012 summary of control activities from Wildlife Services.

In Lower Green Bay, Lake Michigan, cormorants nest on two islands and totaled about 2,100 pairs when management was initiated in 2006. Fishery concerns revolve around potential impacts on yellow perch, which declined by 90 percent between 1980 and 2002. Poor recruitment was identified as the likely factor behind the decline, but commercial fishers and anglers identified concerns about potential impacts from cormorants, whose numbers had increased substantially during the period when perch declined. In 2003, the largest year-class of yellow perch in twenty-five years was documented after more than a decade of extremely poor reproduction; that finding prompted the Department of Natural Resources to initiate a cormorant diet and impact study in 2004. As part of the work, Sarah Meadows, a graduate student pursuing a master's degree at the University of Wisconsin, assessed the cormorant diet over a three-year period. Her results indicated that cormorants consumed a variety of species, with yellow perch, walleye, and round goby important in the diet during the years studied. Yellow perch were a significant part of the diet only when they were abundant and in a preferred size range.[18] However, no estimates were obtained for the standing fish stock in Lower Green Bay, and so the impact of cormorant predation on the fish community in general and yellow perch in particular could not be determined. But while the diet study was under way, cormorant numbers reached all-time highs in the area; simultaneously, yellow perch began exhibiting a substantial recovery. In 2006, angler bag limits were raised, and by 2007 quotas for commercial fishermen had been increased. Nevertheless, the agencies identified the theoretical *possibility* that cormorant foraging may have reduced the magnitude of the yellow perch increase as a reason to reduce cormorant numbers. Changes in vegetation and potential impacts on other waterbirds were identified as additional management concerns.

Therefore, the Department of Natural Resources established a population objective of 1,000 pairs for Lower Green Bay, requiring the elimination of 2,960 breeding birds. Goals included reducing "demands on food resources" and nest space for other birds.

In northern Green Bay, Door County, Lake Michigan, cormorants nest on five principal islands. They totaled about 12,500 pairs when management was initiated. Concerns centered on declines in the sport fish brown trout and rainbow trout (also nonnative), potential impacts on overall fish populations, and the health of the fishery ecosystem in general. In relation to declines in brown trout, the evidence implicating cormorants was that while declines were observed in Lake Michigan waters away from cormorant concentrations, sharper declines were observed in waters closer to cormorant concentrations. Similar observations were reported for rainbow trout. Because these fish stay near shore, agencies speculated they were vulnerable to cormorants throughout the breeding and migration seasons. For potential impacts on the fish community in general, the agencies estimated the potential numbers of cormorants feeding based on counts of breeding cormorants, assumptions about numbers of offspring, and estimates of nonbreeders that ranged from 0.6 to 4.0 birds per pair. Incorporating the upper range undoubtedly resulted in a significant overestimate of cormorant numbers, and the calculations produced shockingly large consumption estimates that suggested cormorants might consume more than sixteen million pounds of fish annually. Meanwhile no data were presented in the environmental assessment on standing fish stocks that would put any level of predation into a meaningful context. As in Lower Green Bay, concerns about vegetation loss were also identified. The Wisconsin Department of Natural Resources established a population objective of 5,000 pairs for the region, requiring a reduction of 14,000 breeding birds. The objective was based on the number of cormorants present in Door County before steep declines in harvest rates for brown trout occurred, and corresponded to a time when concerns about impacts on vegetation were relatively low. Again, goals included reducing demands on food resources and nesting space, with more specific goals of reducing foraging pressure on brown trout and rainbow trout.

For both management units, the agencies noted that resource protection goals related to birds at several large colonies did not require immediate reductions in cormorant numbers. Therefore, a slower and more conservative approach using egg oiling rather than shooting was initially implemented. This more cautious approach would enable agencies to monitor affected resources and adjust management actions gradually in response to new information. But two years after the management plan was finalized, more funding became available to USDA Wisconsin Wildlife Services, presenting the opportunity to reach cormorant population goals more quickly through lethal reduction. The funding led to new strategies being defined to assess vegetation recovery, prevent birds from using other islands, and to evaluate improvements and factors affecting survival in brown trout.[19] As a result, on each of three islands previously identified for the "slow, conservative" approach, 1,000 birds were targeted for culling. In the spring and summer of 2011, intensive shooting was employed in addition to egg oiling, and more than 2,500 cormorants were shot on or around these islands. More progress toward population objectives would have been made, but some islands where birds were nesting are national wildlife refuge lands, and as of 2012 refuge managers had not given permission for cormorant control (see chapter 16). To monitor the fishery response to management, a similar approach to the one used in Michigan was taken here. Creel and trawl surveys are ongoing, and fishery changes will be correlated with changes in cormorant numbers.

Minnesota

Most cormorant control for sport fisheries in Minnesota has focused on Leech Lake, where cormorants nest on Little Pelican Island. When control was initiated in 2005, the colony totaled about 2,500 pairs. Fishery concerns revolved around declines in yellow perch and walleye, and were linked to cormorants through circumstantial data, including increases in cormorant numbers coincident with fishery declines; a correlation between the foraging area of cormorants and the area of the lake where fish declines occurred; and physical similarities between Leech Lake and Oneida Lake, where declines in the same species due to cormorants have already been described. But dramatic declines and comparable low points in the walleye and yellow perch populations were also observed in numerous years when cormorant colonies were not present on the lake. Notably, cormorants did not nest on the lake in any significant numbers until 2000, while a declining trend in walleye year-class strength began in the mid-1990s. Despite those data, and no information on cormorant diet composition, an aggressive culling program was implemented in 2005. No information was available about the level of cormorant predation that could be sustained, and therefore an estimate of sustainable foraging intensity calculated for Oneida Lake was extrapolated to Leech Lake. The exercise identified a population objective of 500 pairs, requiring an 80 percent reduction of breeding birds, or 4,000 individuals. It was thought that this objective would benefit common terns as well as the fisheries.[20]

Detailed diet studies initiated along with management have indicated that in the years studied, the cormorant diet was dominated by yellow perch but included very few walleye, ranging from 0.5 to 3.3 percent of the number of fish eaten. The great majority of walleye and yellow perch consumed were small or young-of-year fish, suggesting that most cormorant predation constituted compensatory mortality. But ongoing diet studies and fish population modeling indicate that the predation on one-plus and older walleye has greater potential to reduce fish populations. During the first year of management, monitoring efforts indicated a dramatic increase in the number of young walleye, one that far exceeded the increase expected from cormorant control alone, suggesting that other factors were influencing recovery. Additionally, even though the breeding population target for cormorants has been reached in most years, the presence of subadult and nonbreeding birds has resulted in cormorant numbers on the lake that are about twice the level of that predicted necessary for fish recovery to occur. Despite the higher number of cormorants, both yellow perch and walleye have been recovering well, and high recruitment and excellent growth rates were observed for walleye in most years. A 2011 report prepared by the Leech Lake Band of Ojibwe to summarize control activities states that this recovery "naturally calls into question to what level cormorants need to be controlled on Leech Lake to allow fish populations to recover and be maintained." In fact, by 2009 or earlier it appeared that a much larger number of cormorants could be maintained while fishery objectives were being met, but no refinements to the population objective have been made.[21]

Evaluating the effectiveness of cormorant management in Minnesota has been impossible because cormorant control coincided with other actions to facilitate walleye recovery, including extensive walleye stocking, reduced fishing pressure, the initiation of a slot limit to protect specific size ranges of fish, and a reduced bag limit. Nevertheless, to a large extent the public has attributed walleye recovery to cormorant management. For example, in 2008 the Department of Natural Resources regional fisheries manager Henry Drewes reported to the media that due to the apparent walleye recovery many people were satisfied with a cormorant population size of 500 pairs, and some preferred that cormorants be eliminated from the lake altogether. Continued research is planned to try to determine the number of cormorants that can be supported on the lake, but it will likely be a complex undertaking, and clear answers may be difficult to obtain. Additionally, as long as funds remain available for cormorant control, public perception and the attitudes of motivated stakeholders will continue to exert a powerful influence on population objectives for cormorants.[22]

Reviewing the information from these programs reveals some broad and disturbing trends among natural resource agencies. First, with the publication of just a few studies indicating that cormorants could negatively impact fisheries in some circumstances, concern about *potential* cormorant impacts increased dramatically. The studies were essentially viewed as "proof" that cormorants are destructive to fisheries, and their results have been cited time and again by agencies seeking to justify cormorant management in areas where similar data do not exist. With such proof on board and adaptive management in place, agencies have bypassed the need for rigorous scientific data when making decisions on how to manage cormorants in most locations. Instead, circumstantial data, extrapolation, and guesswork, bolstered by back-of-the-envelope calculations, are cobbled together to make a case to manage these birds. Frequently, the remotest possibility that cormorants may be having an effect, or could have some kind of an effect in the future, provides sufficient justification for eliminating large numbers of cormorants regionally.

Second, because cormorants are only one of many possible factors that may affect fisheries, and because environmental conditions and ecological processes are themselves evolving, the key requirement for an appropriate monitoring program is not an easy objective to meet. Most management evaluations do not address the issue of nonstationarity or the influence of changing environmental conditions on resource dynamics. Additionally, the agencies state that it may take a period of several years to determine whether management is having the desired effect. But as demonstrated in numerous locations, environmental changes occurring over relatively short time periods preclude or at least complicate the ability to determine how much recovery is due to cormorant control rather than other factors, such as shifts in the composition of the fish community or improvements in fish reproduction. Moreover, achieving such very general objectives as "improvements in fish communities" and "increased biomass" requires a dramatic response from multiple species. But such changes are unlikely to occur over any time period that can be carefully monitored given funding and technical limitations, and during which other environmental changes won't influence fishery resources. Casting yet more doubt on the ability to fulfill the monitoring requirement is the agencies' own admission that the necessary monitoring may be challenging to accomplish, less than perfect, and subject to "very limited" funding.[23] These observations suggest that the type of monitoring necessary for a knowledge-based solution may be impossible to achieve, especially in a rapidly changing environment.

Third, in most cases, the scale of management under way appears grossly excessive relative to the scale of the resource problem. Nowhere is this more apparent than in Michigan, which warrants especially close scrutiny. There, the "experiment" targets the vast majority of breeding birds, and some kind of "treatment" is in place at most locations. But are such large reductions at so many locations really necessary to address potential impacts on fish populations? The state claims that they are, offering two main arguments to justify control at that scale. First, Michigan frequently emphasizes that it has the largest breeding population of cormorants in the United States. However, Michigan is almost entirely surrounded by Great Lakes waters. Connecting channels, including the St. Mary's and Detroit Rivers and Lake St. Clair, also adjoin the state. Michigan has more than 3,200 miles of freshwater coastline, greatly exceeding that of any other Great Lakes state, putting it on par with the coastal states of Texas, California, and Maine. The state also has an abundance of inland lakes, and is comparable in this resource to Minnesota and Wisconsin. These features provide significant cormorant resources in the form of breeding colony sites and abundant fish concentrations, which explains the large numbers of cormorants that occur here. Given the extent of the resources that Michigan possesses, its proportionally larger number of cormorants does not by itself indicate a resource problem.[24]

The state's other justification revolves around the statistical limitations and variation associated with monitoring techniques. In the 2011 environmental assessment for Michigan, the agencies stated that "the change in cormorant numbers needs to be sufficiently large enough to detect a change in the fish population measures given their variance." But that argument does not address why all major breeding areas are being aggressively targeted simultaneously, nor why, in most cases, the largest or nearly largest reductions possible are taken at the outset. Smaller reductions of, say, 50 percent, achieved over a more gradual time period, would still be sufficiently large to provide necessary information and manage potential impacts in many areas. Furthermore, smaller-scale experimental work incorporating a paired design, with sites matched by size, fish community, and water quality, would be far more informative. In this kind of study, cormorants would be managed on one set of experimental waters and not managed on a similar set. Additionally, extensive monitoring on both sets of waters before, during, and after the experiment would be required. To date, only one experiment with this kind of design has been initiated, and it was in Canadian waters to evaluate egg oiling.[25] Results detailing the fish response have not yet been published.

Finally, a fourth disturbing trend is the concentration of intensive management on cormorants as a matter of resource allocation; cormorants are one of the few potential influences on fisheries that can be managed, and relatively easily at that. Agencies have acknowledged that they can control only some of the factors that affect Great Lakes fisheries, noting in this context that the question then becomes whether managing the factors they can address will suffice to overcome the problems faced by the resources they want to enhance.[26] This approach means that cormorants are heavily targeted while other important influences go unaddressed, and a need for ongoing cormorant control becomes a permanent feature of the model. A perfect example of the flawed thinking behind this approach is management targeting cormorants for brown trout. In the Great Lakes, this species is plagued by inferior genetics of hatchery fish, lack of forage fish, competition from other exotic species, the recovery of walleye, and an overall limited niche. Almost nothing can be done about any of

Brown trout

these problems, and the agencies noted that cormorant impacts on a healthy, relatively abundant, and naturally reproducing fish population wouldn't be the same as impacts on a much more limited population of stocked trout.[27] This observation, along with the use of harvest goals for brown trout as the basis for cormorant population objectives, indicates that significant cormorant control will be a crucial part of the "life support" system required to maintain a struggling, nonnative species. In this way, the cormorant becomes the recipient of an extremely unbalanced approach to fisheries management, emerging once again as a scapegoat.

The problems identified here suggest that the general approach to cormorant control is an abuse of the adaptive management concept. The seemingly unlimited cormorant control occurring in the Great Lakes region greatly exceeds what is needed for both hypothesis testing and existing resource problems. Most project goals and data-collection methods are too vague and too limited to adequately assess the effects of cormorant management on sport and commercial fish harvests. Furthermore, despite evidence from several areas suggesting that greater numbers of cormorants could be sustainable and that less aggressive approaches are warranted, management has been unvarying. In fact, none of the programs touted as adaptive have included any adaptation whatsoever, and thus ignore a key criterion of adaptive management. The only "refinements" have involved increasingly aggressive actions, limited solely by funding and accessibility to birds. Ultimately, the view of cormorant predation as a resource allocation issue appears to give the agencies carte blanche for almost unlimited "experimental" reductions. In some cases, the consumption of even one fish of a particular species by the occasional cormorant is deemed detrimental. In others, management is advocated when other data contradict a need for it or even when cormorant predation is beneficial, as when they eat large amounts of round goby. Considered in concert, these problems suggest that adaptive management has been invoked to legitimize a large-scale killing program for cormorants that has little to do with science or knowledge-based solutions.

14

Back to the Wintering Grounds

Liberties with Science and Policy

What may begin as honest error, however, has a way of evolving through almost imperceptible steps from self-delusion to fraud.
—Robert L. Park, *Voodoo Science* (2000)

TRAVELING SOUTH TO the wintering grounds, the cormorant encounters a substantial change in climate, geography, and menu—and a fate sealed by even more potent forces than those operating on the breeding grounds. In this landscape, it is not just recreation and lifestyle that are at stake, but big business, a context that has made the autumnal arrival of hundreds of thousands of cormorants a particularly unwelcome event. Indeed, personal investment, a sense of being robbed, and a struggling industry with very narrow profit margins take the cormorant issue to a whole new level. From Texas to Alabama, both depredation orders are in full swing, and enough cormorants can't be shot.

Perhaps because the stakes are higher, and the tolerance is lower, there has been far less pretense south of the Mason-Dixon line about scientific interpretation, the framing of cormorant conflicts, and cormorant killing. For

On the wing

example, during the 1990s, Arkansas, Oklahoma, and Texas officially declared the cormorant a "nuisance" or "pest" through one means or another,
sending federal authorities a clear message about this federally protected bird.
In 2003, the Arkansas Game and Fish Commission asserted its position even
more strongly and passed a minute order declaring a "no tolerance" policy
for breeding cormorants in the state. The policy was established at the request
of anglers and fish farmers who believed that cormorants were harming both
the state's commercial fish farming and sportfishing industries. Although the
minute order cannot supersede federal law, it did inform the state director of
Wildlife Services that the commission supported and encouraged Wildlife Services' cormorant control efforts in the state. It also announced to the USFWS
that the commission expected Wildlife Services to destroy all nests in the state,
which, as of 2012, the agency continued to do.[1]

Assessing Impacts to Aquaculture

Undoubtedly, cormorants consume substantial numbers of pond-raised fish, leading to significant losses for aquaculture producers. But in any given year, some ponds experience little or no predation while others suffer big losses. The USDA's *Catfish 2010* report documented that 46 percent of food-size operations and 58 percent of grow-out ponds experienced no losses to predation by birds or other animals in the year surveyed, while 5 percent of operations experienced severe losses.[2] Whether operations experience predation problems from cormorants is largely influenced by how birds and resources are distributed. Typically, cormorants do not occur uniformly throughout a given area, but instead are clumped in response to the distribution of resources available. Some birds may prey on natural fish communities, while others key in on artificial resources. And specific characteristics at ponds may make fish more vulnerable at some locations than at others.

To estimate impacts on ponds, the same factors that influence fisheries in open-water systems have to be addressed. Cormorant predation relative to other mortality sources must be considered, but even within these simple systems that can be a complex business. The primary factor limiting catfish production is infectious disease, and outbreaks are not uncommon even on well-run facilities. A variety of viral, bacterial, and fungal diseases, as well as ills caused by parasites and protozoa, constitute major sources of loss. The emergence of disease as a limiting factor coincided with steadily increasing stocking and feeding rates, and with the multi-batch cropping system, which stocks new populations of fingerlings into ponds with existing populations of larger fish. Following in importance is predation by a variety of animals, and since disease and predation are often linked, that interaction adds another element of complexity. Oxygen stress and water-quality issues also constitute significant sources of mortality. Typically, producers don't know the extent to which each of these factors affects their fish. For example, almost three-fourths of the fingerling operations included in the USDA's *Catfish 2010* assessment reported losses to unknown factors, and nearly 40 percent attributed all fry or fingerling loss to unknown causes.[3]

Factors affecting fish growth rates likewise have to be understood. The time required for fish to reach harvestable size is influenced by stocking densities and mortality factors, and as the density of fish in a rearing pond decreases from mortality, growth rates of the remaining fish are expected to increase. At harvest, this is reflected in the size distribution of fish: the fewer fish remaining in a pond may be larger on average than those in more crowded ponds. If fish are valued only by total weight harvested, size does not matter; however, larger-sized individuals may be worth more per pound than smaller ones. In this context, assessing economic impacts of cormorant predation becomes even more complex.

From Experimental Pond to the Delta Region of Mississippi: The $25 Million Example

As described earlier, initial efforts to estimate losses in the Delta region of Mississippi got under way in the late 1980s and involved producer surveys, observations at ponds, and, later, bioenergetics modeling. In the late 1990s, Wildlife Services undertook a nationwide survey of catfish producers about wildlife-caused losses they experienced in 1996. Producers estimated their losses to wildlife at $12 million, and reported spending more than $5 million on wildlife damage management. Combined, producer estimates of loss and damage control equated to about 4 percent of total catfish sales in 1996. During the same time period, the number of wintering cormorants increased in the region, and Wildlife Services biologists used bioenergetics modeling to produce a revised estimate for annual fingerling replacement costs that equated to $5 million annually.[4]

But all the estimates produced thus far fell short in one way or another. The extent to which producers can estimate their losses was considered questionable, and bioenergetics modeling still did not estimate the losses that producers experience at harvest. To further study losses to cormorant predation, in the late 1990s, Jim Glahn and his colleague Brian Dorr initiated trials with captive cormorants at the National Wildlife Research Center at Starkville, Mississippi. The facility, which is completely enclosed with chain-link fencing and netting, includes three small catfish ponds about a tenth of an acre in size and approximately three feet deep. After initial methods were worked out, each pond was divided in half, and each half was densely stocked with catfish fingerlings based on a commercial pond-stocking rate. Additionally, a buffer prey was stocked to help simulate cormorant diet composition in the field. Fish were regularly fed, and oxygen levels were carefully monitored. Cormorants were allowed access to half of the divided ponds for a specified time. After the cormorant foraging period ended, fish were maintained in all pond halves until they reached harvestable size, and their numbers were then inventoried.

Results published in 2002 indicated that the use of alternative prey appeared to reduce the impact of cormorant predation on catfish; nevertheless, cormorant predation led to a 30 percent decline in catfish numbers and a 20 percent decline in catfish biomass at harvest. Predation on fingerlings led to losses at harvest that appeared additive, but the greater growth of catfish surviving predation partially compensated for the early loss of fingerlings. Additionally, predation on fingerlings had variable effects on catfish biomass depending on the circumstances in which it occurred. For instance, disease-related mortality was observed during trials, and in ponds where significant outbreaks occurred, cormorant predation had little or no effect on biomass at harvest; where outbreaks were more moderate predation effect was more significant.[5]

To illustrate the practical applications of their research, Glahn and Dorr extrapolated a 30 percent decline in fingerlings to a typical fifteen-acre commercial pond used in a single-batch production system. This exercise equated to a loss of approximately 22,000 fish with a replacement value of approximately $2,200. They extrapolated the corresponding 20 percent decline in biomass at harvest to the commercial pond scale, which indicated the consumption of fingerlings led to much greater losses at harvest—$10,500, or about five times the value of lost fingerlings. Importantly, the authors acknowledged that several elements arising from the experimental setting might have affected the predation levels they observed. These included cormorants being kept in captivity, and the substantial differences in scale and management between the research ponds and commercial ponds. Additionally, the size of the alternative prey provided, the low mortality of fish, the lack of other predators, social facilitation between cormorants affecting prey choices, and prey densities likely affected predation rates. And although the study looked at impacts in a single-batch production system in which fish of uniform size and age are grown together, most operations in the Delta use a multi-batch production system, in which fish of various sizes and ages are cultured in a single pond, and in which a true inventory of fish surviving to harvest is nearly impossible to obtain.

While all these factors limited the applicability of the findings to the commercial pond scale, that restriction did not prevent the results of the trials from becoming the definitive measure of cormorant damage to aquaculture.

In August 2000, seventeen months before the results were published in a peer-reviewed journal, Glahn and his colleagues gave the previously mentioned, highly influential presentation in which the captive trial results were used for the first time to indicate the potential magnitude of the cormorant predation problem. By assuming that the total replacement costs of fingerlings consumed equaled $5 million, and then multiplying replacement costs by five to estimate losses at harvest, Glahn reported that losses experienced by catfish farmers in the Delta region of Mississippi may equal $25 million. He provided calculations based on the derived production-loss values to estimate profit losses. Using the fifteen-acre commercial pond scale under a single-batch cropping system as the budget unit, he predicted that profits at ponds experiencing cormorant predation would decline by 111 percent.

Needless to say, Glahn's extrapolation catapulted the magnitude of cormorant predation to an unprecedented level. Two years after the presentation, Glahn and his coauthors published an economic consideration of cormorant impacts that incorporated all the figures included in the presentation. Conspicuously absent was any discussion of the limitations of the underlying data associated with the captive trials.[6] Glahn's presentation and its publication are particularly significant because since the presentation was given in 2000, the $25 million figure and the 5:1 ratio underlying it have been continually promulgated in the scientific literature, management documents, and outreach material made available to the public. Within a month of the presentation, Wildlife Services' "science-based initiative" advocating flyway-level management for cormorants (described in chapter 9) became the first publication to cite and use the as-yet unpublished experimental results. In this document, the 5:1 ratio of loss at harvest to fingerling replacement costs is presented as evidence of the magnitude of cormorant predation on southern aquaculture. The initiative, too, neglected any discussion of the numerous issues that might limit the applicability of results from the experimental pond to a commercial-scale operation.[7]

Soon, citations of the estimates that Glahn and colleagues derived from the captive trials emerged in numerous publications and materials addressing cormorant issues. Not surprisingly, the estimate provided justification for the expanded Aquaculture Depredation Order, and was incorporated in both the 2003 final environmental impact statement for cormorants and the final rule for the Public Resource Depredation Order. In 2004, outreach material provided by Wildlife Services on the economic impacts of bird predation on aquaculture stated: "Industry costs associated with bird damage and damage prevention are estimated to exceed $25 million annually." In June of that year, the chairman of the National Aquaculture Association's Bird Depredation Committee used the estimate in testimony given to the US House Committee on Natural Resources. This testimony was presented in support of HR 3320, which had been introduced in 2003 as part of a continuing effort to give Wildlife Services duplicate authority to manage migratory birds. The chairman stated that cormorant-caused "production losses to Mississippi Delta catfish farmers amounts to $25 million or 8.6 percent of catfish sales per year."[8]

In 2005, the estimate made its way into the National Sea Grant Law Center's summary of the lawsuit brought against the USFWS and Wildlife Services for issuing the second depredation order. The opening sentence reveals that the estimate had become a defining feature of the cormorant: "A water bird native to North America, the double-crested cormorant is a fish-eating bird that has been responsible for the loss of at least $25 million in annual catfish production, mainly in the Mississippi Delta." In 2006, the Cornell University Cooperative Extension prepared a guide for the public on cormorant issues and management, and used the estimate more conservatively, stating that losses to the catfish industry ranged from $5 million to $25 million. The range apparently reflected whether replacement costs or losses at harvest were estimated.[9]

In 2009, as the depredation orders were set to expire, the USFWS prepared an environmental assessment to consider the effect of extending them. Since publication of the FEIS in 2003, no new studies had been undertaken to update information on the cormorant's economic impacts on aquaculture. Therefore, the 2009 environmental assessment relied on information presented in the 2003 FEIS, referring the reader to the 2003 document for a discussion of economic impacts associated with cormorants at fish farms. Thus, the $25 million estimate continued to provide justification for cormorant management.[10]

In 2010, another flyway-level plan for cormorant management was recommended to the USFWS for implementation. This one focused on the Atlantic and Mississippi Flyways, pathways that together cover the eastern half of the continent. The plan was prepared by the Atlantic and Mississippi Flyway Councils, which consist of representatives from all federal, state, and provincial agencies with management responsibility for migratory birds in the flyway. Like the "initiative" developed by Wildlife Services, the plan would shift cormorant management from the local to the flyway scale and, if implemented, could have major impacts on cormorant populations. Once again, the $25 million dollar loss estimate, along with the 100 percent losses in producer profits, were cited to justify control at the flyway scale.[11]

Finally, a Wildlife Services fact sheet prepared in 2011 indicates that the agency was still communicating the canonical estimate to the public. But the figure had moved from the hypothetical to the factual. The fact sheet states: "The catfish aquaculture industry experiences nearly $25 million in losses annually due to cormorants alone."[12]

The history of the $25 million estimate provides unique insights into how science has been used and how the cormorant has been portrayed. The limited results obtained from the three tiny experimental ponds came to characterize the cormorant problem across the entire Delta region, and the potentially extreme impacts that cormorants could have on aquaculture became fact. Over and over again, the problem has been presented as one of drastic economic proportions, with farmers severely victimized by this rapacious bird. The $25 million figure has been cited repeatedly to demonstrate the need for aggressive management, and its repetition has made the cormorant guilty of large-scale crime. While cormorant predation might in fact have resulted in damage at that scale, or at an even greater scale, it was equally possible that damages would have occurred at a far smaller scale. But the necessary data to verify the actual level of impact were simply never obtained. What is more, the factors explaining the limited applicability of the captive trial results to the commercial pond scale had long since disappeared into the ether. All that remains is an extrapolation by which the cormorant has become a criminal of epic proportion.

Additional progress toward refining loss estimates occurred when a new study undertaken by Wildlife Services was published in 2012. The research relied on data obtained in the field and at commercial ponds to examine cormorant predation at specific pond types (such as those for fingerlings or food fish). Information was incorporated on cormorant abundance and distribution, along with data indicating that cormorants foraged at fingerling and food-fish ponds in proportion to the ponds' occurrence in the landscape. Estimates of cormorant consumption of fish were calculated by using bioenergetics modeling. The study took into account dietary differences between birds roosting in the interior of the region, where they consume more catfish, and birds roosting near the river, where they consume less catfish. Based on the percent availability of each pond type in the Delta region, the total average biomass of catfish consumed within each pond type was estimated. Results indicated that cormorants foraged most frequently at food-fish ponds, the most abundant pond type, and that losses there were much greater than losses at fingerling ponds. Additionally, loss varied significantly between years. For instance, during the winter of 2000–2001, combined losses estimated at both pond types equaled $12 million, while during the winter of 2003–2004 combined losses equaled $5.6 million. Those losses represented 2.3–4.6 percent of Mississippi's catfish sales. Numerous other factors were identified that can strongly influence economic loss estimates at a given level of depredation. These included volatility in variable production costs, the nominal sale price of catfish, the distribution of depredation on pond types, and the cormorant distribution in interior or river regions.[13]

While the authors identified many factors that could increase or decrease loss estimates, they concluded by stating that their data provided "the most accurate estimate of economic loss to the catfish aquaculture industry in the Yazoo Basin [Mississippi Delta] to date," a statement that has been made about more than one such estimate. Notably, the data indicated that cormorant-caused losses in the two winters examined were much smaller than $25 million, and that the amount of loss could vary significantly between years. Such findings emphasize the need for caution when interpreting and communicating results, especially when results are based on very particular and limited sets of circumstances and have major implications for the public perception and management of cormorants. As seen above, the extrapolation of results can help escalate a problem, set into motion aggressive management policies for cormorants, and steer a course that inevitably leads to large-scale population reductions.

The most recent work by Wildlife Services represents an important step toward more accurate estimates of cormorant-caused losses, but ultimately, more important than the actual damages attributed to cormorants are the environmental practices that created the problem, and continue unabated. Although the problem has been constantly framed by estimates of economic loss, and the cormorant has emerged as public enemy number one, at the root of the conflict is a much larger issue: the transformation of a floodplain along major migratory flyways into densely stocked, unprotected fishponds. Remarkably, catfish aquaculture has been in business for close to fifty years, and for most of that time producers have experienced losses to cormorants and a host of other fish-eating birds and animals. Yet industry practices to significantly minimize predation have been very limited, a fact attributed time and again to the large costs required to bird-proof farms, and to the narrow profit margins of producers. While the responsibility to protect ponds lies with producers, the distribution and abundance of cormorants is repeatedly highlighted as the problem, and the responsibility for losses is attributed to the birds. Currently, night-roost dispersal is still considered the most effective strategy for minimizing catfish depredation by cormorants, a strategy that revolves around managing cormorants rather than ponds. Other important strategies, such as shooting and scaring, likewise concentrate on the birds rather than the ponds. Moreover, no analyses have been undertaken to determine whether it would be more economical to compensate fish farmers for losses either through direct payment or by helping them pursue alternative strategies that might minimize losses (for example, the modification of farms, providing alternative feeding areas for cormorants).

The recent work by Wildlife Services provides information that may help producers focus on the pond types most needing protection and on when to step up efforts to minimize losses. Additionally, it identifies modifications of cultural practices and new technologies for pond-level management that may help reduce losses. But the authors continue to emphasize cormorant management as a key to addressing the problem. While night-roost dispersal is acknowledged as the most effective strategy available, limitations in its effectiveness and implementation have become apparent. Continual disturbance of cormorants at their roosts, combined with increased numbers of cormorants in the region, has led to a dramatic increase in the number of night roosts used since 1990. As a result, flyway-level control continues to be identified as a strategy for preventing losses to the catfish industry.[14]

Relative to the logic of a flyway-level approach, it is important to consider that cormorants using southeastern aquaculture ponds come from across the continent, and an essentially continuous supply of birds is at hand to consume resources. Catfish farms represent a rich and easy food source, and because they are such prime foraging areas, they would be one of the last places where cormorant numbers would decline, even if continental numbers overall were declining rapidly. This suggests that if flyway-level control is to minimize losses at southeastern catfish farms, very large numbers of cormorants will need to be destroyed elsewhere to reduce the number wintering in the Delta. Data on the great cormorant in Bavaria, Germany, support this observation. In the late 1990s, Bavaria implemented an intensive shooting program for great cormorants in order to reduce overall fish depredation. Although the number of great cormorants shot during this period equaled between 50 and 100 percent of the average Bavarian winter cormorant population, the average number of wintering cormorants did not substantially decrease, a result attributed to dead birds being rapidly replaced by newly arriving individuals.[15]

Furthermore, the flyway approach would effectively transfer the responsibility to protect ponds from the producer to the government. That result would, in all likelihood, remove the incentive to modify production practices and manage ponds so that they become less vulnerable to cormorant predation—in effect, perpetuating the problem. Significant funds would be required to support such an effort, and as long as catfish continue to be produced using current culture practices, management would need to be ongoing.

Alternatively, if the cormorant problem for southeastern aquaculture was acknowledged as an environmental rather than an economic one, large-scale population control would not represent a solution. Instead, the cormorant would emerge as a symptom rather than the disease itself. From this perspective, the unnatural concentration of catfish in the Delta and elsewhere in the South would be identified as the source of many widespread or flyway-level problems, such as large increases of cormorants on the breeding grounds as a result of unnatural food supplies on the wintering grounds; large numbers of cormorants and other birds being killed on the wintering grounds; large numbers of roost sites being colonized by cormorants on the wintering grounds because of continual harassment and the dispersal of birds from existing roost sites; and poor environmental practices leading not only to large losses for catfish producers but also substantial problems elsewhere.

The introduction of Asian carp into midwestern rivers and lakes is a prime example of the dangerous and wide reach of environmental practices used by the aquaculture industry. Three of the four most problematic Asian carp species in US waters, the bighead, black, and silver carp, were first imported into the United States in the early 1970s by a private fish farmer in Arkansas for use as a biological control agent and as food fish. During floods, imported fish escaped from ponds, hatcheries, and research facilities. The carp then made their way into several river systems, including the Mississippi, Missouri, Illinois, and Ohio. These accidental introductions have resulted in self-sustaining populations that currently pose numerous large-scale ecological problems in many waters; nevertheless, southern fish producers continue to culture Asian carp.[16]

Looking at cormorant-fishery conflicts from an environmental perspective would ultimately identify aquaculture, rather than cormorants, as the source of the problem, and of many others as well. Economic estimates of fish lost to birds would be but one measure of poor environmental practices, and solutions would be directed at fish-culture methods instead of reductions in birds. But economics and short-term gain have thus far trumped environmentalism and long-term solutions. Hence, an ever-widening arsenal of weapons continues to take aim at the wrong target, the cormorant.

The Special Case of Texas

Texas takes the most liberal approach to managing cormorants for fisheries, revealing yet more disturbing trends among natural resource agencies in regard to this species. Based on data submitted annually to the USFWS by the Texas Parks and Wildlife Department, Texas has been conducting a large-scale cormorant control program since 2005. Between 2008 and 2011, the number of counties where cormorants were killed ranged between 52 and 64 annually, mainly across the southern, central-eastern, and eastern portion of the state. Because this is a state rather than federal effort, no environmental assessment has analyzed potential environmental impacts. Furthermore, the state has not developed any formal cormorant management plan that uses scientific information from specific sites or areas to document the nature and extent of conflicts. The program makes no pretense of being an experiment or an adaptive-management approach. Instead, Texas Parks and Wildlife simply issues nuisance permits for killing cormorants to any landowner who possesses a valid Texas hunting license and pays the current fee of $13.[17]

As described in chapter 10, this unique solution to human-cormorant conflicts is acknowledged by the USFWS as a state-agency control effort authorized under the second depredation order. Permit holders are designated as "agents" of Texas Parks and Wildlife, which makes the agency responsible for the cormorants killed under the permit and for ensuring that cormorant control complies with the terms of the depredation order. Technically, the agency has the responsibility to demonstrate how this control benefits public resources, and must submit annual reports to the USFWS describing aspects of its cormorant management activities. In turn, the USFWS is responsible for making sure the agency is abiding by the regulations in the depredation order.

A Texas Parks and Wildlife report summarizing the agency's control activities from September 2010 through August 2011 defined two management objectives: "to provide the best possible fishing" and to simultaneously protect and enhance the state's freshwater aquatic resources. No direct or site-specific evidence demonstrating impacts or potential impacts was provided; instead, the report simply noted that numerous studies had documented the potential for cormorants to impact fish populations. More specific to Texas, the report included the observation that sport fish can be common in cormorant diets when the birds feed in reservoirs with high fish densities. In such environments, when cormorant abundance is high predation has the potential to collapse some fish populations and affect the overall abundance of several fish species. Because cormorant numbers have increased in Texas, the agency noted a greater potential for such effects to occur in more parts of the state.

These observations were based on a published analysis involving simulations of cormorants feeding at reservoirs in Oklahoma. But the overall results of that study suggested that cormorant predation has only a minor or inconsequential effect on the recreational fishery of a typical reservoir in Oklahoma. Severe effects were found only when high cormorant abundances existed at reservoirs from autumn through spring, a condition not encountered on most Oklahoma reservoirs. Additionally, a 1993 study assessing cormorants' diet in Texas reservoirs indicated that rough fish species were the most common prey consumed by the birds. On some reservoirs, sport fish were a substantial portion of the diet by weight, but not by number. Furthermore, cormorants consumed very few large fish; the sport fish consumed were typically much smaller than those taken by anglers. The authors concluded that overall consumption of desired sport fish in Texas reservoirs was an insignificant portion of the cormorant diet, and indicated that cormorants might even have a beneficial effect in some circumstances by exerting control on the numbers of forage and rough fish species.[18]

Comments by Dave Terre, division chief of Texas Parks and Wildlife's Inland Fisheries Division, provide additional perspective. In 2010, Terre was quoted in *Bassmaster* magazine: "We believe there can be localized impacts on reservoir fisheries. . . . However, because of the seasonal nature of their [cormorant] time in Texas and the fact they do prey primarily on rough and forage fish, the impacts on sportfish are usually not severe, and they are hard to quantify." Further, he noted that on big public reservoirs with good prey populations, Texas has not experienced declines in fisheries, despite cormorants being present at nearly all of them. For these reasons (and somewhat ironically), Texas does not issue cormorant control permits for public waters.[19]

In describing how cormorant control alleviated or was expected to alleviate resource impacts, the Texas 2010–2011 report simply asserted that by controlling and reducing cormorant numbers, "the rate of increase in their population seen in recent years should be slowed. Therefore, the impacts to fish populations and fishing-related industries in the State of Texas should be alleviated, especially in localized areas where cormorant impacts have been severe enough to warrant control measures." But the report concludes by noting that cormorants are an abundant winter resident in Texas, and thus the numbers taken by permit holders is a very small percentage of the tens of thousands of birds that occur in the state.

Significantly absent are data and discussion linking the birds shot on private lands to those potentially impacting public resources, the fundamental link for which this control was authorized under the second depredation order. In fact, the comments from Terre, along with the fact that Texas does not undertake cormorant control on public waters, suggest that cormorants are not posing a threat to public fish resources. No specific studies to monitor cormorant numbers or their impacts, or to evaluate the effectiveness of control in local areas, are mentioned in the report. In the absence of those kinds of data, the effort appears to be based entirely on public perception about what the birds are doing, and on misguided assumptions about how haphazard shooting may alleviate the problem. A review of data submitted for 2008–2011 indicates that in each year, fewer than thirty cormorants were killed, on average, each month in the great majority of counties (84–98 percent). Since tens of thousands of birds winter in Texas, removing that few birds at the county level would be meaningless for minimizing impacts on fish populations in public waters. In a few counties, large numbers of birds have been removed, mostly by individual landowners. To illustrate, 4,199 cormorants were killed in Harris County in 2010–2011, accounting for approximately 70 percent of the birds killed in the state; most of these killings are attributed to one ambitious permit holder. While individual depredation permits have always been available to private property owners, they do not authorize an open season on targeted animals. Rather, they have much stricter requirements and limitations than the nuisance

permit and are more expensive to obtain, factors that, before the second depredation order, resulted in other avenues being pursued when managing personal
problems with cormorants.

The lack of any site-specific data to demonstrate problems or to indicate what level of cormorant control might provide benefits to public resources
raises the question of what possible benefit cormorant control so conducted
might bring, other than to make a few people feel better. As practiced, such
control will not result in changes in cormorant numbers that will lead to benefits on public waters. It will, however, perpetuate the cormorant as a villain.
The program reveals that Texas Parks and Wildlife has openly abandoned ecological assessment as a criterion for wildlife management, at least in regard to
cormorants. Moreover, the acceptance of the program as a legitimate approach
under the second depredation order indicates that the USFWS operates by the
same criteria and appears to recognize no bounds to agency discretion.

15

Engineer or Destroyer

The Case of the Catastrophic Ecosystem Flip

A FEW MONTHS before culling was initiated on Lake Erie's Middle Island by Parks Canada, two member organizations of Cormorant Defenders International requested that a federal court grant an injunction to stop the impending cull. They charged that the statutory requirement to file a management plan for the park had not been completed before the decision to manage cormorants was made. The lawyer for Parks Canada admitted that there was no management plan for Middle Island but argued that if the cormorant population was not drastically reduced, the impacts on vegetation could result in a "catastrophic ecosystem flip." The judge agreed that there was a serious issue to be reviewed by the court, but, fearing the impending "catastrophic flip," allowed the 2008 cull to proceed.[1]

The notion that the cormorant is an agent of catastrophic ecosystem change identifies another thread of irony running through the cormorant's tale. Given the truly devastating impacts that humans have had on aquatic and terrestrial ecosystems, the cormorant's effects are meager indeed. For instance, consider the effect of human overfishing on freshwater and marine fishes. Aquatic ecosystems have been affected in their entirety; profound shifts in species composition have already been described for the Great Lakes, Lake Winnipegosis, Lake La Biche, and the St. Lawrence estuary. For human impacts

Nesting associates, the great blue heron and double-crested cormorant

on vegetation, it is worth noting that the estimated forest area lost globally from human activities was an estimated 32 million acres a year between 2000 and 2010, with the annual net loss of forests estimated at 12.4 million acres a year during this time frame. By comparison, the total amount of combined island area occupied by nesting cormorants in the US Great Lakes between 1977 and 2007 was about 6,100 acres. Moreover, many of the birds using the island spaces were nesting on rock, sand, or bare soil. Nevertheless, the belief that catastrophic ecosystem changes will result from cormorant activities has become an important theme in cormorant management. Between 2004 and 2011, approximately 40,000 cormorants were killed at five Great Lakes sites (High Bluff and Middle islands, Ontario, and three sites in Ohio) for this reason alone, and such management has occurred or is also occurring in New York, Vermont, Wisconsin, Arkansas, and Alabama.[2]

Cormorant control for what is termed "biodiversity impact" contrasts sharply with control to increase fishery allocations. First, the science to assess physical changes arising from cormorant activity at specific locations is far less complex than that required to untangle the underwater mysteries that shape fish populations. The physical changes resulting from cormorant nesting and roosting activities are far more obvious and measurable, and can be directly attributed to the birds. Losses in forest canopy and tree death, along with changes in groundcover, soil chemistry, colonial waterbird numbers, and nest distribution, are some of the easily measured indicators related to cormorant activity. Second, cormorant management to enhance fisheries is undertaken only to benefit humans; conversely, agencies presenting ecological arguments for cormorant control to protect island communities emphasize that protecting habitat from cormorants is a biological necessity that benefits other species and ecosystems. Often, the uniqueness of a particular island, its provision for an uncommon or limited natural resource, or its representative plants and animals are presented in support of this kind of management. Cormorant control for this purpose thus emerges as a more virtuous undertaking than control for fisheries, despite the larger context of drastic ecosystem change resulting from human activity, and is likely to gain the support of a large and diverse segment of the public concerned about the environment.

The dramatic transformations that cormorants can impose on an island have led to an image of the bird as a fundamentally destructive force. In this regard, the elements of time and human memory have not been in the cormorant's favor. The bird's absence or suppressed numbers across much of its range for most of the twentieth century has resulted in specific expectations of what islands should look like without abundant cormorant populations. As the cormorant has recovered, its interaction with the environment has transformed it into a kind of anomaly in the natural world. In many areas, the cormorant is now regarded as an invasive species, one that aggressively establishes itself at the expense of native species and natural ecosystem functions. Invasive species are often exotics that do not naturally occur in a landscape; the cormorant, however, is both native and recovering, and its interactions with the environment are natural. But its historical absence, along with the rapid and expansive way in which it has been able to reestablish itself, has served to highlight its perception as an invasive. Combined with the cormorant's unshakable reputation as a devourer of fish, the image of destroyer has come to define this bird and has led to even less public tolerance.

The cormorant has recovered only to return to a world where most resources have long been spoken for. The environmental assessment for cormorant management in Ohio articulates this development as one of limited wildlife habitat. The document acknowledges that habitat changes arising from waterbird use "would not have been a problem" historically, at least not for waterbird communities, because once habitat at particular sites changed, birds could simply move on to new locations. But in present-day Ohio, human land use has limited the number of alternative sites that can support colonial waterbirds, and many of the sites being used are considered to be at either biological or social carrying capacity (numbers tolerable to humans). In this case, cormorant-induced change *is* problematic, since by transforming the limited habitat available, it may affect other birds and interfere with human desires to see other birds and forests in a particular landscape. The relation between cormorant impacts and historical or current conditions is important because it demonstrates that change resulting from cormorants is not inherently negative, but becomes so only within a particular set of circumstances.

Relative to this last point, cormorant management presented as a biological necessity or as a matter of ecological restoration has especially important implications for perceptions of the cormorant's role in the ecosystem and its place in the landscape. Implicit in such management is the idea that environmental change resulting from cormorant activities is negative. The final environmental impact statement prepared by the USFWS, the state environmental assessments prepared by Wildlife Services, and the management plans for cormorants in Canada all address the natural history aspects that predispose this

species to be seen as negatively affecting its environment. Typical concerns revolve around the bird's potential to diminish biological resources or processes
at particular sites, and in so doing to degrade the environment.

The way in which many natural resource agencies frame cormorant impacts on island resources can be illustrated by two specific and related definitions provided in the Middle Island Conservation Plan prepared by Point Pelee
National Park and Parks Canada. The plan focuses exclusively on cormorant
management intended to restore "ecological integrity," defined in the Canada
National Parks Act (2000), with respect to a park, as "a condition that is determined to be characteristic of its natural region and likely to persist, including
abiotic components and the composition and abundance of native species and
biological communities, rates of change and supporting processes." A definition is provided also for species "hyperabundance": "native species in a national park can be defined as hyperabundant when their numbers clearly exceed
the upper range of natural variability that is characteristic of the ecosystem,
and where there is a demonstrated long-term negative impact of ecological
integrity." In both definitions, a notion of what should be considered "characteristic" of a natural environment is closely tied to expectations for population
sizes of native species. In relation to the cormorant's diminished status in the
past and its great abundance in the present, both definitions inevitably lead to
current populations of cormorants being considered unrepresentative of natural conditions and thus indicative of a serious ecological problem.[3]

Using these criteria, Parks Canada identified the cormorant as hyperabundant and therefore as a "significant and ongoing threat to the ecological
integrity of Middle Island." Similarly, the area used by nesting cormorants at
Presqu'ile Provincial Park was described as a "degraded environment," and
large reductions in cormorant numbers were "deemed necessary to protect
woodland habitat," indicating that cormorants were hyperabundant there too.
Likewise, the Ohio environmental assessment reports observations of (or the
potential for) vegetation damage at the state's five cormorant colonies. Therefore, large reductions in the number of cormorants nesting on the islands and
foraging around them during migration are presented as necessary to protect
vegetation and wildlife—which clearly indicates that cormorants in Ohio are
considered hyperabundant.[4]

To what extent are these categorizations warranted? Considering the
relatively straightforward documentation of island change induced by cormorants, and the fact that habitat for other birds may be limited, or that islands

may support rare communities, these categorizations may have some validity. But a broader consideration of past and present influences on affected island communities suggests that the situation is more complicated than it seems and that documentation of island change by itself is not enough to warrant negative judgments of cormorants. For the degree to which habitat modification, no matter how conspicuous, constitutes an impact is not obvious. Rather, like assessing impacts to fish, assessing impacts to biodiversity requires that they be considered within the broader context of other factors affecting the community. And because the desire for "characteristic representation" is closely tied to expectations of how many birds should be in the landscape, it is essential to understand the cormorant's history in a particular region. An examination of these matters suggests that characterizations of hyperabundance are often misleading and perpetuate the idea that large concentrations of cormorants are an aberration on the landscape.

Island Dynamics

Islands, the location for nearly all cormorant management deemed ecologically necessary, are highly dynamic landforms with characteristic features that *favor the potential for rapid change*. In particular, because islands are relatively small, isolated landforms surrounded by water, with a 360-degree exposure to the elements, they are highly vulnerable to both natural processes and human activities. Islands are extremely sensitive to the forces of erosion and accretion, so in the course of a year they can change dramatically in shape and size. If they are small or low-lying, their vegetative cover and colonial waterbird composition can change substantially over a year, too. Severe storms can significantly alter island morphology and, along with other factors, vegetative cover. Conversely, these same forces can create new islands or reveal old ones that were previously submerged. Seemingly overnight, islands can become part of the mainland or be reduced to a reef. In the Great Lakes, fluctuating water levels have led to well-documented cycles of vegetation loss and reestablishment, along with cycles of use and disuse by colonial waterbirds.[5]

There are many striking examples of just how quickly and significantly islands and their waterbird communities can change. One of the largest colonies of black-crowned night-herons in the Great Lakes once lived on Wolfe Island in eastern Lake Ontario. In 1954, a record noted that 20,000 night-herons were there at one time. But elevated water levels in 1955 killed the vegetation in which the birds nested. Several years of low water levels followed, and the shrubs began to regrow. By 1960, large numbers of night-herons were once again using the site, but later the colony permanently abandoned the island because of human use. More recently, in 2010, a colony of ring-billed gulls comprised by nearly 23,000 pairs abandoned a site at the Lake Calumet dike in Illinois after a number of years of use. Failure to nest at the site was thought to relate to warm spring temperatures, which favored rapid vegetation growth beyond the height suitable for nesting.[6]

While it can appear shocking or disturbing to humans, island colonization by cormorants that results in vegetation and species composition change is only one of many natural processes that can dramatically alter an island's appearance and biotic community. But considered at a broad geographic or landscape scale, cormorant-induced change on an individual island, especially on the small ones typically occupied by cormorants, is insignificant. For example, the Great Lakes are a major breeding location within the double-crested cormorant's range. But as of 2010, cormorants were known to have used just 260 islands over the forty-year period since their population began recovering, representing less than 1 percent of the total islands in the lakes. Furthermore, on the islands used, cormorants often occupied only a fraction of the space available, and many of the islands were not vegetated.[7] Additionally, on some islands where vegetation change and shifts in species composition have occurred, historical data indicate that the vegetation or habitat affected by cormorants was far from pristine.

Considered over the long term, island change due to cormorants is likely insignificant. As recently as the mid-twentieth century, a cycle of forestation and deforestation, in the absence of cormorants, had been observed on many Great Lakes islands. Some became reforested during the period when cormorants were artificially absent or suppressed to very low numbers, only to be changed again when cormorants began to recover. Given the dynamic nature of islands, many changes in their morphology, flora, and fauna have undoubtedly occurred, and more will assuredly take place. Ultimately, of much greater importance than the history of individual islands is the fact that the processes that created historical diversity have long been altered or lost. Exotic species (more than 160 have been documented in Great Lakes waters), pollution, global climate change, ecological regime shifts (of which the cormorant is one), and many other forces now at work are the important factors and processes shaping biodiversity on these islands today.[8] Within this context, managing cormorants in order to prevent island change and to maintain particular communities in perpetuity is a little like building on a floodplain. The dynamic nature of islands and their natural propensity toward change may make such management an exercise in futility.

What Constitutes an Impact?

Negative impacts on individual trees, plants, and birds clearly occur when cormorant activity leads to tree or plant death, or the loss of a nesting effort. But are such impacts significant enough to warrant management action? Typically, impacts at the level of the individual are insignificant at the ecosystem scale. It is only when impacts have population-level effects that they become important. Therefore, the scale at which ecosystems and populations are defined has important implications for assessing cormorant impacts. Currently, in all cases in which cormorants are being managed in order to prevent specific ecological changes, the geographic extent of the ecosystem and the populations of concern are reduced to the scale of the individual island. But although islands exist in isolation, the biological resources they support typically do not. Instead, resources are for the most part distributed across a wide network of islands. In this way, islands are in fact connected, and together they form a larger ecosystem with dynamic elements that change over time. Therefore, while an ecosystem can be defined at the island scale for certain purposes, that scale is often too small for setting meaningful biological objectives for particular resources.

Canada's High Bluff and Middle Islands provide good examples of this. At both sites, the primary goal of cormorant management is the preservation of vegetation. On High Bluff Island, the vegetation in question is not rare, but the woodlot on the park is a unique attribute comprising mature forest, old trees, and an uncommon species association. In addition, the site has aesthetic value, and is considered significant because mature forests on islands in Lake Ontario are relatively rare. But is this resource biologically meaningful for ecosystem integrity beyond the scale of the individual island? The wooded area consists of less than thirty acres; conversely, examples of mature forest can be found across Canada, as well as in many parts of the Great Lakes. Considered from a larger, ecosystem scale, the importance of the High Bluff Island woodlot is political and cultural, but not ecological.

Middle Island presents a more complex case. Here, preservation of vegetation is directed specifically at Carolinian vegetation, nearly all of which has been destroyed in Canada by deforestation and other human activities. Characterized predominantly by deciduous trees, Canada's remaining Carolinian forest is confined to southwestern Ontario, where it reaches the northern limits of its distribution. Middle Island is one of the few islands in the western basin of Lake Erie with a remnant of this vegetation. But in the eastern United States, which lies literally yards from Middle Island's southern shore, Carolinian vegetation is common and has its greatest representation from the Carolinas (hence the name) to southern New England and eastern Ohio. Looking at information compiled by the USDA, Environment Canada, and other agencies on the status, distribution, habitat needs, and threats to the species of trees, shrubs, and plants identified as concerns on Middle Island, it is not clear how killing cormorants at this location is going to contribute significantly to the conservation of these plant species. Furthermore, since Carolinian vegetation is widely distributed elsewhere, and since Middle Island encompasses just forty-six acres, the loss of this resource from such a small area would be biologically insignificant.

Nevertheless, the Carolinian forests on Lake Erie islands have important conservation value. They are distinct from those found on the mainland, having greater plant diversity relative to the small area they occupy, which makes a case for their preservation.[9] This consideration suggests that the problem with cormorants is a significant one. But the existence of a real problem does not mean that the cormorant is appropriately categorized as hyperabundant or as a threat to ecological integrity. This is an important distinction, because framing the issue in that way perpetuates the myth of the cormorant as a destructive force uncharacteristic of natural environments. In fact, the real conflict is not complex: cormorants are destroying trees that are valued by humans in this particular island environment, but are common elsewhere. Many animals damage trees in their day-to-day activities, and as noted earlier, humans destroy trees hourly all over the world. But these activities typically constitute an ecological impact only when large tracts of forest disappear. Why then do the natural and small-scale activities of cormorants get moved into the realm of significant ecological impact?

In assessing cormorant impacts on other colonial waterbirds, the individual island scale is more obviously problematic. Waterbird populations are not typically confined to individual islands, but instead use a number of sites throughout a geographic region at any given time. Since the late 1970s, waterbird populations have been continually monitored through the Great Lakes

Colonial Waterbird Survey, which is coordinated every ten years by the Canadian Wildlife Service and the USFWS. Additionally, particular colonies and specific regions have been sporadically monitored since at least the 1930s. Data from these efforts demonstrate that colonial waterbirds are highly mobile and frequently abandon sites, colonize new sites, and return to abandoned sites after many years of absence. Movement among islands can be influenced by factors such as physical changes in habitat, human disturbance, predators, and changes in food supplies.

Fortuitously, the GLCWS was initiated just as cormorants were reestablishing themselves in the Great Lakes, and the database provides information on colony and population trends for the species typically of concern in relation to cormorants—tree-nesting herons, night-herons, and egrets. Since the survey's initiation, the number of egrets has increased dramatically in the region, while night-herons and great blue herons have experienced some decline. For night-herons, population declines were driven mostly by declines at one site, West Sister Island, Ohio, and were attributed largely to the loss of optimal nesting habitat that occurred through natural forest succession long before cormorants were present. For great blue herons, reasons for declines are not clear, but many colonies were abandoned at locations that were not shared with cormorants.

After the third survey was completed, population data for herons and night-herons in US waters obtained between the late 1970s and the late 1990s were reviewed to assess potential impacts from cormorants. Population declines and abandonments of individual colony sites by both species were common at sites where cormorants did not occur. Close examination of sites where night-herons declined in the presence of cormorants revealed that at nearly all those sites, night-heron declines could not be linked specifically to the presence of cormorants. Other factors that could have influenced the declines included vegetation succession, storms, spatial stratification in nesting areas, and undercounting by surveyors. The study concluded that despite increasing numbers of cormorants, population trends of herons and night-herons did not indicate that cormorants had negatively affected the breeding distribution or productivity of either species regionally. Based on these results, the authors suggested that "cormorant control policy should not be justified by the 'assumption of potential impacts to other waterbird species' without careful documentation of the situation at specific colony sites of concern."[10]

Data from the most recent survey, in the late 2000s, indicated that only great blue herons have continued to decline since the late 1990s. The declines are largely due to the disappearance of two large colonies from US waters, which were not shared with cormorants. While large declines in numbers of night-herons were documented at particular colonies that were shared with cormorants in Canada, the Great Lakes–wide population remained stable overall, suggesting that any potential impacts from cormorants have not re-sulted in regional population declines for this species.

It is also important to consider that herons, night-herons, and egrets have never been particularly abundant in the Great Lakes. These long-legged waders require shallow waters for foraging, and most of the optimal foraging habitat of this type lies inland. Thus, these species are not highly characteristic of Great Lakes waters, at least not compared with cormorants, gulls, and terns. The egret, in fact, has been undergoing a gradual northern range expansion for decades, and likely did not become established as a breeding bird in the Great Lakes until the mid-1940s. Additionally, despite the fact that the Ohio environ-mental assessment claimed that habitat for colonial waterbirds was limited, the great blue heron is widely distributed and common in inland areas of Ohio and other Great Lakes states and provinces. The night-heron is also expanding its range northward.

Like Carolinian vegetation, most heron, night-heron, and egret popu-lations are found outside the Great Lakes, and it is not clear how much any of these birds move between coastal and inland areas. For these species, the Great Lakes population may not represent the true population, because the survey area is not based on known distributional boundaries, but on the geographic area that can be reasonably monitored given survey limitations and logistics. This area includes the five Great Lakes, their connecting waters, and shoreline up to half a mile inland. Since birds that abandon Great Lakes islands may move more than that far inland, a decline in the Great Lakes population may not necessarily represent a decline in the "true" breeding population. For all of these reasons, the extent to which actions to diminish cormorants in the Great Lakes will provide significant benefits for populations of these species is unknown.

Last, conservation science indicates that impacts to great blue herons and egrets in the region may not be significant at a meaningful biological scale, which is much larger than that of the individual island. The Upper Mississippi

Valley and Great Lakes region is one of sixteen North American planning areas for which regional waterbird conservation plans have been developed. The plan for this region identifies great blue herons and great egrets as "not at risk," a status assigned to species whose populations are stable or increasing and for which there is no major conservation concern. This is the same status assigned to the double-crested cormorant. While the plan recommends preservation of colony sites for herons and egrets, the recommendation should be balanced with the low conservation priority assigned these species. On the other hand, the night-heron was designated as a moderate-to-high conservation concern, which gives the species a higher priority for conservation action and suggests that cormorant impacts on it have a greater potential to be biologically significant. Therefore, in addition to the preservation of breeding sites, cormorant management on a site-by-site basis was recommended as a conservation strategy if negative impacts from cormorants were documented. But notably, the plan acknowledged that while "scientific documentation of cormorant impacts is a standard beyond the Public Resource Depredation Order," it advocated "waterbird management actions be developed with a sound scientific foundation." Specifically, the plan recommended that scientific analyses first document whether significant cormorant impacts are occurring as a precursor to management and to aid in the identification of cormorant population levels required to reach acceptable levels of impacts on resources. Such evaluation would logically include an assessment of a particular site's importance to a species' biologically defined population, information that should be weighed against the magnitude of the cormorant management that may be deemed necessary to preserve it.[11]

None of this is to suggest that individual islands do not have conservation value. As noted for Middle Island, distinctive ecological components make a case for individual island preservation. Likewise, Lake Erie's West Sister Island, where more than 10,000 cormorants have been killed since 2006, hosts a unique waterbird colony. The fourth Great Lakes Colonial Waterbird Survey (2007–2010) indicated that the island supports proportionally large numbers of Great Lakes night-herons, great blue herons, and great egrets. It also has a substantial history of use by these birds. Furthermore, it is a beautiful forested island that is aesthetically pleasing to look at. Despite these features, however, and for the reasons identified above, cormorant management on even this island does not appear necessary at the regional population level for these species. Rather, its benefits are island specific, preserving the aesthetic quality of the island for people and fostering the persistence of a particular group of birds. Within that context, it is important to acknowledge that cormorant management undertaken for biodiversity may be more about what people want and expect to see on the landscape than about protecting ecological integrity.

The Importance of History

Perhaps the most compelling challenge to claims that the cormorant is over-abundant and threatens ecological integrity is plain history. For most of the twentieth century, cormorant populations were suppressed to unnaturally low levels or were absent from particular areas. Early accounts of the Great Lakes document how, upon discovery of breeding colonies, fishermen immediately began to persecute cormorants and suppress their numbers. As early as the mid-1940s, when the population size was estimated at between 1,000 and 3,000 pairs, governmental efforts to reduce cormorant numbers were initiated in Ontario, indicating that cormorants were considered hyperabundant even at that small population size. Not long after, cormorants were devastated by contaminants. Thus, cormorants did not begin to demonstrate what natural numbers in the region might look like until the contaminants were diminished and the birds obtained protected status in the 1970s. But by then, new food sources were present on the breeding and wintering grounds, leading many to believe that the number of cormorants that developed was unnatural. This may or may not be the case, but at present there is no way to determine whether current numbers "clearly exceed the upper range of natural variability that is characteristic of the ecosystem," a criterion for evaluating hyperabundance. Since monitoring programs got under way, cormorants have not been present in a natural state long enough for researchers to be able to determine what the upper range of natural variability in a particular ecosystem might be, especially the Great Lakes.

The only baseline for comparison is the population size between the 1920s and 1950s, when cormorant numbers were small and their distribution was limited. But because of known human influences, these numbers cannot be considered characteristic of the Great Lakes ecosystem. Nevertheless, they apparently underlie expectations about cormorant abundance and related ecological conditions. In many of the current management plans for cormorants, population objectives are based on numbers of birds present before the effects of cormorant abundance became obvious or before strong concerns developed about the birds' potential to affect resources. These numbers, mostly reached during the late 1980s and early 1990s, still do not identify an upper range of natural variability characteristic of the ecosystem. Instead, they identify a population size that begins to exceed human tolerance limits.

Considering the cormorant's natural ability to transform landscapes, and the historical suppression of its numbers, fundamental questions arise about what may or may not be representative of an ecosystem. For example, does the landscape that evolved in the cormorant's absence represent what really characterizes the environment, or is *it* actually the aberration, representing degraded ecological conditions? This is one of the arguments articulated by activists opposed to management in Canada. To explore this possibility, the history of islands and colonial waterbird populations in western Lake Erie over the last 160 years is particularly informative.

Based on historical records, the biological community on these islands today appears substantially different from what it was a century or more ago. Visits to islands in the eastern half of the western basin in the early 1900s indicated that nine of the ten species that currently nest there were not known to do so at the time. Only the common tern bred widely in the area. Breeding records for the others are available from about 1900 on. Numerous factors influenced colonial waterbird occurrence at that time. Populations of many birds, particularly herons, egrets, cormorants, gulls, and terns, were diminished by hunting for the millinery trade, egging, and persecution. In Ohio, cormorants were regular migrants along western Lake Erie in the 1800s, and had been observed around Sandusky Bay in summer in the late 1870s. They were also known to breed inland at a few locations south and southwest of Sandusky Bay. But by the turn of the century they had been extirpated from the state and were becoming uncommon even as migrants. Similarly, great blue herons disappeared from portions of the state, and egrets, which may have been common late-summer visitors, became rare during the same period.[12]

Conversely, night-herons were not heavily persecuted, and their Ohio breeding population underwent a range expansion between 1915 and 1935. It is believed that they became established along western Lake Erie by about 1920, and one of the first islands they colonized was West Sister. The island may have been particularly attractive at that time because at least half of it was in a grass-shrub stage, which provides favorable nesting habitat for this species. But these conditions were a direct result of human activity. During the mid-1800s, a lighthouse was erected, and the keepers' quarters were built on the island. Many trees were cut for firewood, the land was grazed by domestic livestock, and even domestic turkeys were raised. In the late 1930s, the lighthouse became automated and people moved off the island. At some point during this period, the island was colonized by great blue herons, which by the mid-1930s were recovering substantially in Ohio and elsewhere. By 1937, the island was recognized for its significant night-heron and heron populations and designated as a National Wildlife Refuge.[13]

At about the same time, the first breeding records for cormorants were reported for western Lake Erie, on two small islands in the vicinity of sites currently occupied by breeding cormorants today. The first breeding records for great egrets in western Lake Erie were also obtained. Like the great blue heron, egrets had begun demonstrating a substantial recovery in Ohio by the

Great egret

1930s, and by 1946, had colonized West Sister Island. The breeding records from this period indicate an expanding breeding range and they are, in fact, the first breeding records for both Ohio and the Great Lakes.[14]

Over the next few decades, great blue herons, night-herons, and egrets continued to increase in number and to colonize additional sites in western Lake Erie. But persecution by fishermen, legal control in Ontario, and the effects of contaminants on reproductive success continued to suppress cormorant abundance and distribution in the Great Lakes. In the western basin of Lake Erie, cormorants persisted in very small numbers at just one site, Big Chicken Island. By 1970, the entire Great Lakes population consisted of fewer than 100 pairs, with almost half at the lone Lake Erie site. Finally, toward the late 1970s, cormorants entered their recovery phase, and in the early 1980s they began colonizing islands in western Lake Erie that were already occupied by the other tree-nesting waterbirds. Given the increases and expansion in the populations of other colonial waterbirds between the 1940s and 1970s, it is likely that cormorants would have naturally expanded during this period also if they had not been subjected to persecution and contaminants. Had they done so, current communities on numerous islands may have been very different from what they are today. This history supports the idea that the cormorant's absence, rather than its presence, is the true indicator of degraded ecological conditions.[15]

A further point: on many of the islands with ecological conditions that agencies hope to maintain, those conditions are not necessarily representative of the natural region. In fact, just the opposite may be true. For instance, Middle Island was inhabited to some extent from the 1800s through the mid-twentieth century, and human activity resulted in significant island alteration. A channel was dredged to provide a protected area for boats; grapes were grown for wine production; buildings were erected, including a hotel with electricity, fireplaces, and a casino that allegedly drew up to 200 visitors a day; and an airstrip was constructed for small private planes. In 1976, the island was purchased by a private owner and has since been allowed to return to a more natural state. Because of the heavy human presence, it was not colonized by either tree-nesting herons or cormorants until the 1980s.[16]

Similarly, in addition to the human presence that modified West Sister Island in the mid-1800s, the island was used by the US military during World War II for artillery practice and as a bombing test site. Such activities "degrade" island conditions far more extensively than do naturally occurring native species. At the same time, recovery from such events indicates that islands have a certain resiliency and quickly revert to more natural conditions over time. These historical observations suggest that concepts such as ecological integrity and biodiversity may be invoked to justify cormorant management, but are inappropriately applied.

Finally, specific ecological conditions resulting directly or indirectly from human activity are not likely to persist once islands are again subject to natural forces. The human-related absence of tree-nesting waterbirds from Middle Island throughout most of the twentieth century probably contributed to the persistence of Carolinian vegetation at the site and enabled the island to become and remain densely forested with tall trees. Likewise, once West Sister Island was free from human disturbance, the grass-shrub stage that humans created slowly converted to taller and denser forest, eliminating optimal night-heron habitat, which at this location consists of trees less than fifteen

Black-crowned night-heron

feet tall. For night-herons to persist on the island, intensive management is required to maintain optimal vegetation conditions. Since night-herons likely colonized the site in response to conditions arising from human activity, and since vegetation must be managed in order to maintain night-herons, the site clearly does not represent natural conditions indicative of ecological integrity. Rather, it represents a frozen moment in time. Importantly, vegetation and cormorant management may preserve habitat there for night-herons, but neither that result nor the vegetation conditions maintained should be confused with ecological integrity.

16

Opening Pandora's Box

Some Ethical Implications of Cormorant Management

THE CORMORANT'S TALE has been and remains one of cultural bias, prejudice, and discrimination against creatures that transgress human boundaries. A cultural pariah, the bird is defined, devalued, and excluded by its otherness. The last five hundred years have seen the cormorant transformed first into an exile and later into an exotic without ever leaving its home range. Ironically, in this enlightened age of conservation and heightened environmental awareness, the cormorant's most recent history represents some of the most heinous treatment the bird has received to date. In fact, one of the most remarkable aspects of the double-crested cormorant's story is how it has continued to be treated in modern time, when the lessons of such animals as the passenger pigeon and the wolf are in full view.

The agencies managing cormorants have considered the more than half million birds destroyed since 1998 an inconsequential loss relative to the bird's total population size. And from this perspective, the only one considered, the managing agencies are correct. Well over a million cormorants are estimated to reside annually in North America, and take of 40,000–50,000 birds yearly likely represents less than 5 percent of the population. Clearly, the cormorant

BKM

is currently not in any danger of becoming extinct. But is this the only criterion by which the intense efforts to suppress cormorant numbers should be judged? By any measure, a half million of anything constitutes a large quantity; if visualized, it leaves a trail of dead cormorants lying end-to-end for more than 250 miles. The destruction of so many birds over so short a time has required tens of thousands of cormorants to be targeted yearly. In addition, millions of dollars have been spent to sustain the effort. The sheer magnitude of birds killed, regardless of the population impact, along with the money spent, is certainly not without significance.

As described in earlier chapters, scientific study to document cormorant impacts has not been a required component of the policies that developed for management of the birds in the United States. Nor have most agencies taking aggressive actions to reduce cormorant numbers justified the need for their actions through appropriately obtained scientific information. This is an important oversight because science, the closest thing available to an objective measure of reality, can help lead to rational human behavior and distinguish it from that driven simply by desire or perception. Because cormorant policy has no rigorous requirement for science or the need to distinguish between perception and reality, it is undiscerning and thus fundamentally flawed.

But for policy, science is only part of the equation. The other key element is ethics, a less quantifiable but clearly recognizable force guiding human endeavor. Like science, ethics is a form of practical reasoning that also exerts a strong influence on human behavior and decisions. Unlike science, however, ethics does not reflect objective reality, but rather moral underpinnings, which are ultimately shaped by cultural traditions and other subjective experiences. As a result, a wide variety of perspectives usually develop on controversial issues, along with different opinions on what is right and wrong, and good and bad. This has led to efforts to establish ethical guidelines to govern behavior in a great variety of circumstances, ranging from how children treat each other in the classroom, to formal disciplines such as bioethics and environmental ethics that provide a recognized forum for ethical issues and practical guidance for ethical decisions.

In 2005, the environmental ethicist Ben Minteer and the biologist James Collins published "Ecological Ethics: Building a New Tool Kit for Ecologists and Biodiversity Managers." The paper identifies the lack of any equivalent to bioethics or environmental ethics in the field of ecological research and biodiversity management, and highlights the need for a new approach in practical ethics, one that they term "ecological ethics." Within that framework, they raise numerous ethical questions related to managing wild animals within ecosystems, pointing out that such questions should be considered in management decisions. Nowhere is the need for that sort of ethic more apparent than in the world of cormorant management; in fact, it is the complete lack of any ethical consideration that characterizes the core of the bird's story and its management through time.

In recent policies established for cormorants in the United States, no ethical guidelines have been incorporated to govern decisions about if, when, or how to manage the birds. In the rules outlining regulations for the two depre-dation orders, the word *ethic* never appears. Furthermore, the agencies that de-veloped these policies appear to view their own role in managing cormorants as one that does not engage in questions of ethics. For instance, in the final en-vironmental impact statement, ethics are mentioned in a discussion of the hu-man dimensions of wildlife management, and the existence of varying attitudes about how cormorants should be managed is acknowledged. But the discussion goes no further. Instead, it shifts to the USFWS mission, which is "fish and wildlife biology," with an agency focus on "biologically justified management strategies that are based on the best available science." Similarly, in response to a public comment questioning whether it's right to kill fish-eating birds in particular circumstances, the agencies acknowledged that there are varying attitudes about the ethics of management, but their role is to "help facilitate management of conflicts while ensuring that cormorant populations remain healthy."

Just as scientific evidence has not been a necessary requirement to launch cormorant control, ethics have been even less of a consideration in decisions to manage these birds. But this issue is far more significant, because in the absence of ethics, a door has been opened which leads the way to undermining many of the critical institutions that protect birds and were established only through much blood, sweat and tears.

The Illusion of Protection

In the United States, the two depredation orders provide a great range of circumstances in which large numbers of cormorants can be killed and their nesting efforts destroyed. Of the two, the Public Resource Depredation Or-der constitutes the most important challenge to any sort of protection the bird gained through either the Migratory Bird Treaty Act or the designation of sites as protected areas. In this regard, the impact of the order is particularly signifi-cant in two ways. First, it targets birds allegedly affecting public resources and redefines all the essential elements for life as off-limits to cormorants. Second, to protect these public resources, birds are generally targeted on public rather than privately owned lands. The distinction is important because most of the public lands where birds are targeted are in state or federal ownership precisely because of their importance for the protection and conservation of wildlife. Typically, the lands have been preserved because they are representative of natural communities and support large numbers of colonial waterbirds and other biodiversity. Often, public lands where cormorants are managed have a special conservation or protection status, with official designation as National Wildlife Refuges, Federal Wilderness Areas, and National Parks. In Michigan and Ohio, for example, large numbers of cormorants have been culled or their nests destroyed on five islands designated as part of National Wildlife Refuges, three islands that are part of Federal Wilderness Areas, and one island that is part of a National Park. In Wisconsin, the environmental assessment developed for cormorant management specified reductions on islands that are part of the Green Bay and Gravel Island National Wildlife Refuges, both of which are also Federal Wilderness Areas. Thus far, however, refuge managers here have not allowed cormorant management on these lands.

The original mission of the National Wildlife Refuges was, just as the name implies, to provide refuge for birds and animals that had been intensively hunted or persecuted. Each refuge where cormorant control is under way or has been proposed was specifically established "as a preserve and breeding ground for native birds" or "to protect native and migratory bird habitat." The sites were intended to provide nesting colonial waterbirds a haven from human molestation. Today, this is still largely the refuges' purpose, and land is set aside to enable natural communities to persist out of harm's way. Allowing cormorant control at such sites, particularly when control is based on fishery objectives or otherwise serves no environmentally meaningful purpose, completely undermines the spirit in which the refuge system developed and dismantles the protection it affords.

Controlling cormorants on lands designated as Federal Wilderness Areas is an even more blatant violation of the purpose for which those lands were set aside. In 1964, the Wilderness Act established a national wilderness preservation system in the United States. The system is designed for the protection of areas designated as wilderness and for "the preservation of their wilderness character." The act provides the following definition of *wilderness*:

> An area where the earth and its community of life are untrammeled by man, where man himself is a visitor who does not remain. An area of wilderness is further defined to mean in this Act an area of undeveloped Federal land retaining its primeval character and influence, without permanent improvements or human habitation, which is protected and managed so as to preserve its natural

conditions and which (1) generally appears to have been affected
primarily by the forces of nature, with the imprint of man's work
substantially unnoticeable; (2) has outstanding opportunities for
solitude or a primitive and unconfined type of recreation.[1]

This definition prioritizes lands affected "primarily by the forces of nature,"
as opposed to those affected by humans. The cormorant is one such force of
nature shaping and influencing the primeval character of the land. The act was
developed specifically to set aside lands where such forces could be seen. This
concept has been sidestepped, however, by a policy drafted by the USFWS for
wilderness stewardship. The draft policy establishes a "non-degradation principle" by which the prevailing conditions in an area when it was designated as
wilderness are then used to establish the area's baseline "wilderness values";
the USFWS will not allow those conditions to be degraded. Ironically, the policy strives to maintain a particular set of conditions, despite the Wilderness
Act's emphasis on preserving lands where natural conditions are shaped by
forces of nature.[2]

   Relative to this nondegradation policy, the Ohio environmental assessment points out that West Sister Island was designated as both a National
Wildlife Refuge and a wilderness because of its heron and egret colony, which
implies that cormorants would not be part of the prioritized wilderness values.
By this reasoning, cormorants could probably be excluded as a focal species
from most refuges and wilderness areas, since many of them were designated
at a time when cormorants were in minimal numbers or absent entirely because
of human persecution. This is one more example of how the cormorant is excluded as a reputable member of the waterbird community, and excluded from
the benefits such membership should confer. Ultimately, if cormorants can be
killed even on lands designated as refuges, and aren't allowed to influence the
character of lands specifically identified as wilderness, then there is simply no
place on the landscape where cormorants can exist in natural numbers.

   On private lands, special circumstances further demonstrate that cormorants are essentially safe nowhere, and that the protection afforded by the Migratory Bird Treaty Act is on the way toward being dismantled. To illustrate,
the states of Michigan and Texas once again provide powerful examples. In
Michigan, several islands targeted for management are owned by the Michigan
Nature Association. Established in 1952, this nonprofit organization, which is
committed to protecting special natural areas throughout the state, has pur-

chased several islands in order to provide sanctuaries for birds and wildlife. The organization believes that the justification for cormorant control in Michigan rests on insufficient information, generalizations, and perceptions about cormorants rather than on scientific study. Therefore, it will not allow cormorant control on its islands. But because these islands support nesting birds that agencies believe are consuming too many fish, they remain prime locations for management. To legally kill birds associated with these islands, Wildlife Services defined a buffer zone of 500 yards around each island's shore in which birds would not be shot. But as soon as birds departing from the islands fly beyond the buffer zone, they are legally shot by Wildlife Services personnel waiting in boats just outside the buffer zone. This strategy completely undermines the sanctuary provided by the Michigan Nature Association and pays no regard to the organization's goals. In fact, the strategy uses the sanctuary, a place where cormorants aggregate, to target large numbers of birds. Between 2006 and 2011 (but excluding 2010), this "off-colony shooting" killed well over 11,000 cormorants that were using the islands during the nesting period in the Thunder Bay and Bays de Noc areas.[3]

In Texas, the authorization of nuisance permits under the second depredation order reveals the extent to which the order is open to unorthodox, seemingly perverse interpretation. The order has essentially given states free rein when establishing the rules for cormorant management, and for all intents and purposes, Texas has used it to create a public license-to-kill program. Moreover, by authorizing a policy in which any private landowner can kill unlimited numbers of cormorants year-round after paying a $13 fee and presenting a valid hunting license, the USFWS is complicit in a process that comes another step closer to removing the bird's federal protection.

Conflicts with Conservation Goals

Over the past two decades, large-scale conservation plans for birds have shifted away from a focus on threatened and endangered species to one that is broader and more representative. Some themes are to "keep common birds common" and to link species and population goals to historical distribution and abundance. In 2002, John Fitzpatrick, then the president of the American Ornithologists' Union, published a paper titled "The AOU and Bird Conservation. Recommitment to the Revolution," which recommended a robust mission statement for bird conservation: "Ensure persistence of all American bird populations in their natural numbers, natural habitats, and natural geographic ranges." In addition to specific plans and visions for bird conservation, the Important Bird Area program initiated by BirdLife International has identified thousands of specific sites important for the conservation of birds and biodiversity worldwide.[4]

From the perspective of these bird conservation initiatives, cormorant management is often in direct conflict with conservation goals. For instance, many of the islands in the Great Lakes where cormorants are intensively managed are designated Important Bird Areas. Additionally, many were identified in a separate project undertaken for the USFWS to prioritize the most important breeding sites for colonial waterbird conservation in the US Great Lakes. Designation of these islands as Important Bird Areas or important breeding sites confers no legal protection to birds nesting or roosting on them, nor does it guarantee conservation of the site. But it does indicate that such places have significant conservation value. It stands to reason that killing cormorants and oiling their eggs at sites so designated undermines the site's conservation value, unless it is determined that the site provides unique importance to other birds using it.[5]

The vision of the North American Waterbird Conservation Plan, published in 2002, is one "in which the distribution, diversity, and abundance of populations and habitats of breeding, migratory, and nonbreeding waterbirds are sustained or restored throughout the lands and waters of North America." Yet in many areas where the cormorant has returned to formerly documented breeding sites or areas, it is actively managed. For example, the cormorant historically nested in Arkansas, but was absent for most of the twentieth century. It resumed breeding in the late 1980s at Millwood Lake, but because of conflicts with fishing interests and aquaculture, the colony was controlled by Wildlife Services beginning in 1999. Then, in 2003, the Arkansas Game and Fish Commission declared its "no tolerance" policy for breeding cormorants in the state (see chapter 14). Such a policy of suppression for a native species attempting to reoccupy its former breeding range is in clear opposition to conservation goals designed to restore waterbird populations to their historical range. Furthermore, Millwood Lake is a man-made lake stocked with a giant Florida strain of largemouth bass; ironically, the cormorant is one of the few natural elements present. But the USFWS has not challenged the state's authority under the second depredation order to exclude breeding cormorants from Arkansas, again indicating the agency's unwillingness or inability to enforce the terms of the Migratory Bird Treaty Act.

Similarly, cormorants returned as a breeding species to one of their few formerly documented breeding sites in Ohio, Grand Lake St. Marys, after more than a hundred years of absence. In the late 1860s, cormorants were abundant enough at the site to be described as breeding in "considerable numbers," and for a "boatload" to be killed at one time. They successfully recolonized the site in the late 1990s, but by then great blue herons had been nesting there for perhaps fifteen years. The Ohio environmental assessment identified the lake as important for recreation and walleye fishing, but the reason given for cormorant control there was to prevent impacts on herons. A minimum population objective of fifteen pairs was established for cormorants, which has required killing hundreds of birds. Again, maintaining the only historical colony in the state at fifteen pairs, a size so small that the colony remains at continual risk of extinction, obviously disregards goals to restore populations to their former range. In fact, the agencies state that they "based their management decisions primarily on the current situation and not what may or may not have occurred before the landscape was altered by Europeans." But the "current situation" is that herons are widely distributed and common across the state. Management at this site provides yet another example of action undertaken either because a particular species is favored over another, or worse, that heron conservation is used as an excuse to minimize cormorants.

Finally, cormorants are frequently managed because they are perceived as a threat to conservation goals. But as was described in the previous chapter, little or no distinction has been made between tree loss and island change as caused by cormorants, on the one hand, and true ecological impact on the other. Thus far, managing the cormorant as a biodiversity impact has dispensed with the need to demonstrate just how meaningful these impacts are, and to forgo posing ethical questions about management. Nonetheless, important ethical questions surround the issue: Should ten thousand cormorants be killed in order to save a unique collection of trees on one island? Or to save habitat on one island for other birds that are of low conservation concern and have many other nesting locations? Or to save habitat for birds presenting greater conservation concern but for which there is no compelling data to indicate that cormorant management will be helpful? Do the trees or the other birds have greater value than the cormorants because they occur in more limited numbers? Are social,

284 Science, Management, and Ethics

cultural, and historical priorities enough to justify the destruction of such large numbers of cormorants? Questions of this nature are integral to the more equitable treatment of cormorants and to ethical decisions made about them. As a start, an honest distinction must be made between management undertaken for cultural, political, and social reasons, and management undertaken for truly ecological purposes.

Population Objectives: An Unfair Proposition for Cormorants

Most of the objectives set for cormorants encompass massive reductions or establish populations of such small sizes that they would almost never be naturally maintained except in truly degraded environments. The fact that reductions ranging from 50 to 100 percent are based on little or no scientific data, or simply on the *potential* for impacts, demonstrates the extent to which cormorants have been devalued in North America, and warrants a discussion of the ethical issues inherent in population objectives. To address this aspect, one need only consider that many of the population objectives identified for cormorants have been influenced almost entirely by factors that guarantee the cormorant's unfair treatment. One particularly potent influence has been that of fisheries management, which is exclusively focused on protecting fishery resources so they can be intensely exploited by humans. Given that the USFWS has shifted from a policy of not managing cormorants for fisheries interests to one that allows their management for resource allocation, ethical consideration is even more important. For without it, what limits the number of cormorants removed in fishery allocations to benefit human interests?

Considered within the complex world of predator-prey interactions, the fisheries manager is essentially an advocate of the prey—in this case, fish—with the ultimate goal of making them available to one particular predator, the angler. Needless to say, there is little tolerance for fish lost to cormorants. Fisheries managers, as was made clear in the environmental assessments, view cormorants as one factor in the landscape influencing fish populations that they can potentially limit. This view inevitably leads to unnaturally low population objectives that maximize fisheries interests and minimize those of cormorants, so much so that even poorly adapted exotic species are favored over a well-adapted native one. Such objectives require continual cormorant management or even the extirpation of cormorants from an area, because they have nothing to do with biological carrying capacity or ecological integrity. Rather, they are focused on allocating resources to meet human demands to the fullest extent possible.

Cormorants, unlike humans, must consume fish to survive. Native to North America, they have returned to many portions of their former range and expanded in others. So even in situations where cormorant predation is having well-documented impacts on fisheries, decisions to manage cormorants raise inherently ethical questions. Should cormorants be killed and their nests destroyed simply because they compete with humans? Does this competition require a large-scale management program like that presently employed in the Great Lakes? Should population objectives for cormorants be determined by fisheries managers, who have an obvious bias and essentially work for interest groups and a limited segment of the public? In situations where poor environmental practices have directly created the problem, as in the Delta region of Mississippi, should the cormorant pay for the damages? Or should the catfish industry recognize predation by fish-eating birds as a cost of doing business? Perhaps if it did, better management of fishponds would result, minimizing the extent to which fish-eating birds are a problem in the South and elsewhere.

Another potent factor unfairly influencing population objectives for cormorants is wildlife acceptance capacity, which is often driven by perceived rather than real damages, long-held cultural biases, feelings of entitlement to a particular resource, and sheer hatred. Indeed, the last is so prevalent that it led one journalist to describe the cormorant as "the most hated bird in the world."[6] Intensely negative attitudes, as evidenced by vigilante activity, media reports, and postings on angling forums, remain strong among some segments of the public, and these individuals are typically the most vocal about cormorants. Such attitudes and beliefs predispose this bird to irrational treatment. Moreover, the hatred and misperception that largely characterize the human-cormorant conflict sharply distinguish wildlife acceptance capacities for cormorants from those of many other animals perceived as overabundant, such as white-tailed deer or Canada geese. Because wildlife acceptance capacity does not distinguish between real and perceived impacts—and being hated by humans is such an integral part of the cormorant's past and present—this concept has little utility in defining reasonable and rational population objectives for these birds. In this regard, it must be emphasized that natural resource managers have a responsibility not only to humans but also to the animals, plants, and resources they manage. As a result, the well-known bias of particular segments of the public toward cormorants should preclude wildlife acceptance capacity as a basis for cormorant population objectives. In turn, acknowledgment of this bias should preclude taking a fundamentally biological approach to a problem that is largely social and economic.

Nevertheless, the concept of wildlife acceptance capacity underlies a great deal of cormorant management. In the final rule for the second depredation order, the USFWS states: "We believe that managing certain species to address economic or social concerns . . . is consistent with our mission." But its most extreme manifestation appears in the Atlantic and Mississippi Flyway Plan for cormorant management, submitted to the USFWS in 2010 by the flyway councils. The plan states: "Cormorant numbers have exceeded acceptance capacity with several wildlife stakeholder groups throughout Canada and the United States." That assessment comes from work published by Taylor and Dorr in 2003, who made the statement within a broader discussion identifying the discrepancy between acceptance and biological carrying capacities for cormorants. Taylor and Dorr acknowledged that cormorant numbers are operating under biological carrying capacity; the environment can easily support far greater numbers than people are willing to tolerate, and therein lies the conflict. As a solution, the Atlantic and Mississippi Flyway Plan attempts to establish "tolerable breeding pair population levels" for each state and province within the two flyways, defined as

> the approximate number of double-crested cormorant breeding pairs within a state or province in which the viability of the population would not be impaired (if so desired) and other species and habitats would not be significantly impacted. This definition incorporates the states which choose not to have a viable population of cormorants as well as those which do not feel a need for cormorant management at the present time.

Of the thirty-two states and seven provinces encompassed by the flyways, twenty states and one province defined state- or province-wide population objectives. Seven states established reductions for their populations; Mississippi and Arkansas set the tolerable population level at zero. Six states set population objectives that would maintain current population sizes, all of which had small breeding populations that ranged in size from about 100 to 2,000 pairs. The remaining states and single province established population objectives that would allow for modest increases.[7]

While the plan specifies that states and provinces do not have to enforce the "tolerable levels," many are already doing so or working toward them. Additionally, the development of a "tolerable population level" for a particular state or province gives anti-cormorant segments of the public something to clamor about, and has a high probability of becoming a self-fulfilling prophecy. The concept, and especially the freedom it gives a state or province to eliminate its breeding population or maintain it as a remnant, defies all previous efforts to conserve and protect birds, particularly species that have been historically persecuted. Again, it presents a biological solution to a social or economic problem, and arguably incorporates neither science nor ethics. The irony of tolerating and encouraging the suppression of a native species in its natural range while natural resource managers raise concerns about preserving biodiversity cannot be too highly emphasized. At best, the concept of wildlife acceptance capacity, from which springs the tolerable population level, can be used relative to cormorants to identify those segments of the public that have a particularly low tolerance and why. In most cases, that information could be used to improve public understanding of the biological constraints affecting cormorants, fish, and other natural resources of concern. That approach would be entirely different from the current "shoot first, ask questions later" policy, and would require a high degree of communication, education, and support from natural resource managers, the public, and legislators.

Humane Issues: An Unacknowledged Aspect of Cormorant Management

Observations at High Bluff Island by Canadian activists made it abundantly clear that during a cull many birds succumb slowly to their injuries and experience a drawn-out and painful death. To date, the unique documentation obtained for that location is the only cormorant cull the North American public has been allowed to see. Yet it appears to be fairly representative of what happens to some proportion of birds in all large-scale culls, for even when special efforts are made to minimize humane issues, such as those at Middle Island, some level of wounding is still documented. Nevertheless, in the United States, where most culling occurs, humane issues are not a key concern. The reasons for this are multifaceted, but an important one is that most Americans are simply unaware of the humane issues surrounding cormorants and their control.

In the United States, most culling is conducted by Wildlife Services personnel, who are expert sharpshooters. But because cormorants are targeted under a wide variety of conditions, results similar to those documented on High Bluff Island may be fairly common. For instance, at several locations, Wildlife Services uses a method called pass shooting. With that technique, birds are shot in flight from offshore of islands, and typically experience much higher wounding rates than birds shot on nests. To illustrate, Steve Mortensen, the Fish, Wildlife, and Plant Resources program director for the Division of Resources Management, Leech Lake Band of Ojibwe, generously provided the following observations for the cull at Leech Lake, Minnesota. Wildlife Services uses pass shooting to dispatch about 85 percent of the birds targeted there. Flying to and from feeding areas, birds pass by shooters at various distances, angles, and speeds; it requires on average three to four shots to kill one. If a bird is hit and falling with its head up, shooters keep firing because there is a good chance that the bird will attempt to dive when it hits the water. Wounded birds are chased down with a boat and dispatched, but some inevitably get away. Some are also shot yet able to maintain flight, only to die of their wounds at a later time. Wounding or crippling rates at Leech Lake from pass shooting are thought to be slightly lower than what is reported for waterfowl, which averages about 20 percent based on birds shot with shotguns. But these rates are based on self-reporting by hunters and are out of date; presently there is widespread speculation that waterfowl crippling rates may be twice as high.[8]

The intense disturbance that occurs to the entire colonial waterbird community during a cull presents another humane issue. Typically, human presence in a colony causes adult birds to flush from nests. Eggs are sometimes broken as incubating birds get up hastily, or are eaten, along with small chicks, if gulls are present. Older chicks sometimes fall from tree nests, and mobile chicks on the ground will often run from their nests, only to be attacked later by adult birds whose territory they have wandered into. Because cormorant colonies are typically shared with other waterbird species, each of which has a unique breeding pattern, there is always the potential for cormorant control activities to affect nesting efforts of other birds. Agencies undertake efforts to minimize impacts, but it is difficult to time entry into a colony without compromising the reproductive efforts of at least some birds. At many US locations, visits are made to colonies every few weeks to oil eggs. The reproductive stages of other birds are more vulnerable at some times than others, and those repeat visits make it even more difficult to avoid reproductive losses.

The orphaning of young birds poses yet another humane issue. To minimize this occurrence, a moratorium has been placed on culling in the US after chicks have hatched. But for a number of reasons, it has not been fully effective. First, newly hatched chicks in tree nests are not obvious, and indirect information and signs must be used to surmise their presence. Second, the timing of culls sometimes follows presumed dates for hatching and fledging rather than local colony patterns. Third, shooters may be unaware of chicks and signs of chicks in a colony, or they may choose to shoot anyway. In 2012, colleagues in Michigan visited Gull Island in northern Lake Michigan to take a census of birds two days after Wildlife Services culled cormorants there. During the census, they heard chicks begging loudly, and estimated that the chicks were at least a week old. Similarly, in 2010 researchers were working with birds on Whiskey Island in Lake Michigan, a colony that tends to run a bit later in nesting activities than others. Adult culling began according to presumed fledging dates, but chicks had not yet fledged.[9]

As the agency carrying out most control operations, Wildlife Services has attempted to minimize the importance of these humane issues. In each of the environmental assessments prepared for the Great Lakes states, there is an identical section titled "Humaneness and Animal Welfare Concerns of Methods Used by WS." In it, Wildlife Services acknowledges that "pain obviously occurs in animals" and that "cormorant control methods, especially lethal control, may raise issues about humaneness and animal welfare." But such concerns are bypassed by presenting the issue of pain and suffering as a "complex concept" open to interpretation. The potential experiences of pain and suffering that a cormorant may have during a cull are framed as vague, subjective, and fundamentally disconnected. The agency states: "Wildlife managers and the public would be better served to recognize the complexity of defining suffering since '. . . neither medical nor veterinary curricula explicitly address suffering or its relief.' Therefore . . . humaneness appears to be a person's perception of harm or pain inflicted on an animal, and people may perceive the humaneness of an action differently." The agency adds that "the challenge in coping with this issue is how to achieve the least amount of animal suffering within the constraints imposed by current technology and funding." With this statement, the agency makes clear that external factors beyond its control, rather than its own discretion, ultimately determine the extent to which cormorants will suffer during management activities.[10]

These arguments attempt to remove the matter from the realm of ethics, but a moment's pause to consider honestly what an animal with a brain and a nervous system as highly developed as the cormorant's must experience during control operations centers the issue squarely in the ethical domain. The idea that humane issues arising from cormorant control are matters of human perception and are due to forces outside the agency's control are simply attempts to rationalize the brutal treatment to which this bird has been continually subjected.

Wildlife Services: Another Conflict of Interest

The role and influence of Wildlife Services is another obvious and potent area for ethical consideration in the management of cormorants. The evolution of the agency, particularly since its transfer back to the USDA, has seen it go from one whose function in the past was initially that of a "hired gun" to one that is now far more autonomous. The agency designs, develops, and implements large-scale management programs for numerous wildlife species. With a specific vision "to improve the coexistence of people and wildlife," Wildlife Services is looking more and more like a natural resource agency. But its recent track record indicates that its vision continues to rely on killing *huge* numbers of wild animals yearly. For instance, in 2000, the agency killed 3,267,838 wild animals—birds, mammals, snakes, and turtles. In 2008, the agency destroyed 4,210,411 blackbirds alone.[11]

This destructive approach to resolving human-wildlife conflicts continues to raise questions and garner criticism about the role the agency should play in wildlife management. As recently as 2012, the American Society of Mammalogists submitted a letter to the Legislative and Public Affairs Office of USDA/APHIS to urge the redirection of Wildlife Services management operations and the substantial reduction of funding for lethal control of native wildlife species, especially for native wild mammals. The society stated it found "little evidence that the focus or practices of WS regarding native mammals have changed substantially from its progenitor agencies in the Bureau of Biological Survey 100 years ago" and that the agency's "ongoing record of lethal control stands in stark contrast" to a growing consensus among ecologists "on the integral value of intact ecosystems with their apex predators, and the pervasive ecological damage done by removing them from natural systems." The letter documents some of the controversial issues related to Wildlife Services programs, provides data on the large numbers of mammals killed yearly, and summarizes the society's belief that "current science does not support much of WS's lethal control of native mammals, that it is wasteful and often counterproductive." Another recent example is a lawsuit brought against Wildlife Services in 2012 by WildEarth Guardians, an environmental nongovernmental organization that focuses on the protection and restoration of wild animals, lands, and waters in the American West. The group charges that the agency has relied on an outdated, eighteen-year-old environmental analysis for its activities, and that the document does not consider "new reports on the risks and inefficiencies of its program, evolving public values for wildlife, and new scientific and economic information concerning wildlife management." The organization ultimately wants the court to ban the agency's heavy use of poisons, guns, and traps.[12]

Given all these considerations, should Wildlife Services be the agency responsible for managing conflicts with migratory birds, especially a bird that is hated? The agency has continually advocated a population-level approach to cormorant conflicts and has helped implement lethal management programs in many areas. But rather than providing permanent or sustainable solutions, that approach offers ones that will be ongoing and expensive to maintain. Additionally, it reinforces the belief that cormorants and other fish-eating birds should be killed, and that killing is the right thing to do. And since the agency receives federal dollars to carry out cormorant control, there are strong incentives for the agency to implement widely its "Integrated Wildlife Damage Management" approach for this species.

Soon, one hundred years will have passed since Rosalie Edge developed her brochure about the "misnamed and perverted" Biological Survey, and she would most certainly continue to classify Wildlife Services this way today. During a century of wildlife damage control, the program has relied heavily on population reduction to solve problems. Killing extremely large numbers of wild animals continues to be the agency's defining feature. To be fair, the agency may play an important role in situations where population reductions are truly necessary. Likewise, the National Wildlife Research Center has been a leader in developing nonlethal control methods for wild animals, and has conducted research leading to an improved understanding of cormorants and other fish-eating birds. But the USDA's and Wildlife Services' expertise does not lie in the conservation and management of wild bird populations. More significantly, the agency's track record and philosophy for conflict resolution diminish the likelihood that ethical considerations will be incorporated in decisions to manage cormorants. The agency's primary responsibility to protect agriculture and human interests from wildlife leaves little room for questions like "Should catfish losses to cormorants be considered a cost of doing business?" and "Should fishery resources be primarily allocated to humans?"

It is interesting to consider whether cormorant control under the second depredation order would have evolved into its present form if Wildlife Services did not exist or if the agency's management efforts were significantly different. In the absence of an infrastructure, funding, logistical support and an agency with expertise in population reduction, cormorant management would probably be far more minimal. While vigilante activity might increase, it is debatable whether anglers and other limited segments of the public in today's world would organize to the extent necessary to cover the costs and to regularly visit remote islands in the Great Lakes to shoot cormorants and destroy their eggs. It is also questionable whether such activities on a large scale would be so easily accepted by other segments of the public if they were not legal and occurring under the aegis of the federal government. In considering these hypothetical scenarios, it is worth remembering that in Canada there is no equivalent of Wildlife Services. Yet highly organized illegal activity is not occurring at most locations, despite the country's even larger number of cormorants and similar conflicts with human interests. The one known exception to this may be Manitoba's Lake Winnipegosis, where widespread illegal activity was regularly undertaken at least through the 1990s by commercial fishermen, who may be more capable of undertaking it than other segments of the public. While it is possible that such behavior may develop elsewhere in Canada, in the absence of a governmental infrastructure for cormorant control, biological solutions in the form of population reduction have for the most part not been developed in recent times to address perceived human impacts.

Persecution Masquerading as Management

While the draft and final environmental impact statements for cormorant management were being prepared, the conservation committee of the American Ornithologists' Union formed a panel to review the documents and provide feedback to the USFWS. The panel consisted of five ornithologists with expertise on cormorants and migratory bird management. In 2002 and 2003, the panel filed formal comments on both documents. Additionally, it prepared a formal report that included a review of the perceived problems with double-crested cormorants, the science used to document them, and a thorough critique of the draft and final impact statements for cormorant management. Based on its review, the panel concluded that the impact statements were flawed for the following reasons:

> 1) the scientific evidence supporting the proposed action is weak; 2) the analysis of the data is simplistic; 3) the management plan proposed by USFWS is inadequate and has a poorly evaluated potential to be effective; 4) the consequences of the proposed action on the cormorants are unknown, and appear to be punitive instead of mitigatory; 5) the assessment of success is unclear; in the DEIS, success is based on public perception and not on scientific results. The FEIS is not clear on how success will be assessed; and 6) there is no adequate mechanism for monitoring the population effects of the plan, nor for deciding when to terminate management actions.

The panel also stated that the final impact statement "fails to discriminate effectively between facts and opinions, uses economic arguments without sufficient demonstration of their accuracy, and disregards geographic scale." But the panel's most critical comment by far was that based on its review, "it appears that what the USFWS plans to do constitutes persecution of a bird species rather than a solution to the real problems of declining fisheries and depredation at aquaculture and hatchery sites."[13]

With that statement the panel demonstrated great foresight. For what has emerged over the nine years since the panel wrote those comments has been persecution, plain and simple, masquerading as management. A careful review of the science used to evaluate impacts, along with a consideration of numerous ethical issues, indicates management has proceeded much like a witch hunt. But many features of the cormorant management program conspire to make the euphemism *management* a fairly impenetrable disguise. Not all birds are destroyed, and in most locations at least a small number are left to be "representative" of the environment. The cormorant is usually not openly demonized by managing agencies, and few individuals would discriminate against its black color or refer to it as the "N-bird." Additionally, based on vague ideas about the need to restore "environmental balance" many people believe that agencies are managing resources wisely. Most have no idea that managing cormorants to enhance fishery allocations to humans, even in those places where it might work, is far removed from restoring any natural balance. Nor would they be aware that such management caters to interest groups and is driven by layers of politicking, money, and greed.

Instead, the cormorant is persecuted in more insidious ways. Science is misused, ignored, or stretched to the limits to paint a particular picture. The cormorant's potential to influence several natural resources simultaneously is emphasized in order to create an image of a dangerous animal, and cormorant control is presented as a necessary evil. Economic analyses that consider few relevant factors and are undertaken by an agency with a clear conflict of interest not surprisingly indicate losses in the hundreds of millions of dollars from the adverse effects of cormorants. Those hypothetical costs are then used to justify the expense of control and to emphasize the dire need for it. In management plans lacking the relevant data to identify biologically meaningful objectives, or lacking any good reason to manage cormorants at all, concepts such as "adaptive management," "wildlife acceptance capacity," and "tolerable population level" are invoked in an effort to legitimize a campaign to destroy large numbers of cormorants and allocate large numbers of fish to humans. In reality, the bases for much of this campaign against cormorants have been and continue to be social issues, greed, and irrational human hatred for the birds— not ecology. Despite all we know about cormorants, the same basic reasons for cormorant hatred one hundred years ago are at work today: the cormorant is a trespasser, a radical, and a glutton, and its sleek black plumage continues to inspire fear and loathing. The biggest difference between the cormorant management of today and the cormorant persecution of earlier times is that now cormorant control is carried out mostly by paid wildlife professionals working for the government, rather than just fishermen, and has become a highly systematic and institutionalized practice. As such, it is probably far more effective than the persecution in times gone by.

How exactly has this state of affairs come to pass? The answer lies largely in the USFWS's decision to allow management to occur without scientific justification or ethical considerations, and to pass authority and responsibility for cormorant control to state, tribal, and other federal agencies. The USFWS still has federal oversight for how the bird is managed, but with the emergence of the new and revised depredation orders, particularly the second one, Pandora's

Ready to fly

box was opened and an incredible precedent was set for the management of
fish-eating birds. Bloody paths leading in multiple directions were immediately
established by state, federal, and tribal agencies, and they continue to spread.
Does this transfer of authority and shared burden of responsibility explain why
the USFWS has not said no to the myriad aggressive proposals it received to
destroy cormorants? Did it preclude the agency's authority to say no even to
the most outlandish of them, or even to challenge such obviously inappropriate
campaigns as the "no tolerance" policy waged by Arkansas or the nuisance
permit established by Texas?

While answers to these questions have not been provided here, likely
factors influencing the agency's acquiescence include the very strong lobbies
of the catfish, baitfish, and sportfishing industries. Likewise, there has been a
return to policies of old in which resources are intensely managed for human
allocation and satisfaction, despite the costs to other features of the environment. The management of double-crested cormorants by two federal agencies
serving two very different purposes has also been a significant factor. But despite these other forces, the USFWS's regulatory authority and ultimate responsibility for conserving and protecting migratory birds puts the agency at
the center of the storm. In the end, it is the US Fish and Wildlife Service that
has enabled a level of cormorant destruction that may even exceed the persecution the birds experienced in the nineteenth century. With this agency, then,
lies responsibility for what has arguably become a modern-day policy of persecution for one of nature's most magnificent but misunderstood birds. That
such a policy could arise at the start of the twenty-first century reveals just how
far the United States has yet to go before achieving a resilient ethic of wildlife
management, one that includes even those creatures that lie outside the sphere
of human acceptance.

Afterword

What Future for Cormorants?

WHILE I WAS finalizing this book, the Pacific Flyway Council prepared a plan to address localized conflicts with double-crested cormorants; it involved managing their numbers and distribution at the Pacific Flyway scale. Additionally, the US Army Corps of Engineers began developing an environmental impact statement to manage cormorants in the Columbia River estuary. Although current numbers on the Pacific Coast are probably still below historical levels, the preparation of these documents indicates that cormorants west of the continental divide have gained a similar nuisance status to those in the eastern portion of the range. In the Midwest, cormorant management remained intense: record numbers were killed in Michigan in 2012, and Minnesota expanded its management of cormorants to ostensibly benefit fisheries in 2013. The Great Lakes Restoration Initiative, a federally funded effort for projects addressing the most significant problems in the Great Lakes ecosystem, approved substantial funds for cormorant management. The appropriated funds, which could have been used to address other ecosystem issues, identify the destruction of cormorants as an essential measure of Great Lakes restoration, on par with efforts to clean up toxic substances, combat invasive species, and protect watersheds from polluted runoff. The Great Lakes Fishery Commission began funding control activities, providing an inkling of just what the

297

cormorant is up against. In Congress, Minnesota representative John Kline proposed HR 3074, which would amend the Migratory Bird Treaty Act so that states would have the same authority regarding cormorants as the secretary of the interior. This bill was not passed, but in 2013 it was reproposed by Minnesota representatives Michele Bachmann and Collin Peterson. The legislation was prompted largely by residents in Lake Waconia, just outside the Twin Cities. Many people there believe that cormorants are damaging fishing and the related economy, despite biannual surveys, conducted by the Minnesota Department of Natural Resources, indicating that numbers of game fish in the lake are normal or higher.

On June 30, 2014, the two US depredation orders for cormorants are scheduled to expire, and the USFWS is planning to prepare a supplemental environmental impact statement or environmental assessment to revise regulations for cormorant management. Based on support for, and the intensity of, the management already under way, the depredation orders likely will be extended for another several years. Additionally, the recent developments in the western United States suggest that the reach of the Public Resource Depredation Order may expand to include birds in that portion of the country.

Yet there are also a few glimmers of hope to suggest that strides toward a more enlightened path for human-cormorant interactions are being made. As my book was in the making, two other books on cormorants, one by Dennis Wild and one by Richard King, were also completed. These works tell different stories, but important to each is the extent to which cormorants have been misunderstood and demonized in Western culture. That three such books on cormorants should be published within a two-year span speaks to the sense of injustice that characterizes the cormorant's current treatment by humans. Looking to the scientific literature, the fact that the fisheries scientist James Diana felt the need to challenge publicly the work of a colleague, as well as to pronounce cormorant control on the Great Lakes as an "ethically compromised" undertaking, is also highly significant. Diana's publication, the only one of its kind thus far, provides especially important documentation of the significant scientific and ethical questions that surround cormorant control.

Looking more broadly, the work by Bruce Warburton and Bryan Norton identifying the need for a "knowledge-based ethic" to guide the management of "nuisance" wildlife has important ramifications for cormorant management. These authors recommend that all lethal control operations targeting nuisance wildlife be first reviewed for appropriate experimental design, perhaps by an animal ethics committee—again emphasizing the need for a close look at both the science and ethics behind wildlife control. Moreover, the recent reviews of adaptive management provide important guidelines that, if followed, could provide some much-needed boundaries for cormorant control. And perhaps even more noteworthy is the 2013 publication of a paper in the journal *Conservation Letters* calling for the reform of federal wildlife control in the United States. This paper, coauthored by Bradley Bergstrom and several other university biologists, offers an important scientific perspective and stance on lethal control. The authors conclude that the federal government's heavy reliance on lethal control "is a vestige of the outmoded mentality of western expansionism, in which the goal was to 'tame' the wilderness, replacing the ecosystem's primary-consumer trophic level entirely with domesticated herbivores and a few favored game species and all higher trophic levels with humans. Shortly after this paper's publication, the Center for Biological Diversity, Project Coyote, and the Animal Welfare Institute, with support from the Animal Legal Defense Fund, petitioned the Obama adminstration to reform Wildlife Services. Specifically, these petitioners called for the agency to develop a more ethical, scientific, and transparent approach.

Finally, aggressive management in Canada has remained minimal overall, and with the exception of Alberta, provinces have not undertaken any significant cormorant control for impacts on fisheries. In Ontario, activists have remained steadfast in opposing lethal cormorant control, and some innovative approaches to managing conflicts around biodiversity matters have emerged. At Tommy Thompson Park, strategies to encourage birds to nest on the ground appear to be working, and in 2012 the Toronto and Region Conservation Authority set up a remote webcam to show the public the evolution of the ground-nesting colony over the breeding season. Dramatic photos posted on the website capture a spectacular natural phenomenon in the midst of Toronto's heavily urbanized environment. At Presqu'ile Provincial Park, the approach taken by the conservation authority appears to have been influential, and efforts are now being made at Presqu'ile to deter birds from nesting in particular areas and to attract them to others by enhancing ground-nesting and alternative tree-nesting habitat (namely, platforms). Methods there are still being modified but show promise for alleviating conflicts.

So far, the United States has not looked closely at Canada's alternative approaches and management policies. But with the current restrictions on government funds, the rationality of some of Canada's policies may become more attractive. Moreover, as information becomes available to dispel the myths and misconceptions about these birds and their treatment, understanding, tolerance, and even appreciation for them should increase. This is the greatest future hope for cormorants. Millions of years old, these birds have gathered, fished, and transformed landscapes for eons. But for the past five hundred years, cormorants have been able to exist in North America without being persecuted only when present in very low numbers. For the double-crested variety, such population sizes are unnatural, and will exist without human intervention only when environments are radically altered and fish are impossible to find. Taking the long view, all the way back to *Enaliornis* and those first foot-propelled diving birds that ate fish, humankind would be far better served by simply recognizing the logic of the form. In so doing, it may also recognize that the "solution" represented by destroying cormorants is itself just one more reflection of the real problem at hand: the inability of humans to understand their own role in the vast web of life.

Notes

1. Aristotle's Raven

1. Grémillet et al. 1999.

2. Grémillet et al. 2004.

3. For more detail on feather structure, see Johnsgard 1993 and numerous publications by Grémillet and colleagues at http://www.cefe.cnrs.fr/en/ecologie-spatiale-des-populations/david-gremillet, accessed May 2013.

4. See Strod et al. 2004 for underwater vision changes in great cormorants. Johnsgard 1993 describes adaptations of the eye of the double-crested cormorant.

5. See Grémillet et al. 2005 for polar night diving, and White et al. 2007 for visual acuity.

6. White et al. 2008.

7. Chesser et al. 2010.

8. Van Tets 1976, Siegel-Causey 1988, Siegel-Causey 1999.

9. See *Birds of North America* species accounts.

10. For more information, see Jackson and Jackson 1995 and Wires et al. 2001.

11. See Johnsgard 1993 for the economic value and history of fishing with cormorants.

12. Johnsgard 1993, Von Brandt 2005.

13. Discussion on Hesperornithiform birds summarized from works by Feduccia (1980, 1999).

14. See Brodkorb 1963 and Johnsgard 1993 for Suliform (formerly Pelecaniform) species and Phalacrocoracidae evolution; see van Tets 1976 for Australasian origins.

15. Brodkorb 1963.

16. See Brodkorb 1963 for Pliocene and Pleistocene fossil evidence; Hatch 1982, Hobson and Driver 1989, and Bovy 2011 document evidence from Indian middens.

17. Evolution of *Homo* from the Smithsonian Institution's National Museum of Natural History.

18. Pollard 1977.

19. Arnott 2007.

20. Brachet 1878.

21. Marzluff and Angell 2005 discuss the crow family's "unusually potent cultural staying power."

2. The Double-Crested Cormorant

Note: General information on the natural history, distribution, and diet of the double-crested cormorant is summarized from Johnsgard 1993, Hatch and Weseloh 1999, Siegel-Causey 1999, and Wires et al. 2001, and sources cited therein.

1. Johnsgard 1993; Grémillet 1995; Sellers 1995.

2. Reported in Enstipp et al. 2001.

3. Doucette et al. 2011.

4. See King et al. 1995 for activity budgets, and Stickley et al. 1992 for comment on foraging behavior.

5. Cairns and Kerekes 2000.

6. Cuthbert et al. 2002; Weseloh et al. 2002; Rippey et al. 2002; Hebert et al. 2005.

7. Audubon 1843; Lewis 1929, 24–25; Mendall 1936, 40.

8. Birt et al. 1987.

9. Ibid.

10. Ridgway et al. 2006; Ridgway 2010.

11. Lewis et al. 2001.

12. Personal communication from Brad Allen, Maine Department of Inland Fisheries and Wildlife, and Steve Windels, Voyageurs National Park, Minnesota; for the Pacific Coast, see Adkins and Roby 2010.

13. See Wires and Cuthbert 2010 for the Great Lakes, and Glahn, Reinhold, et al. 2000 for information on wintering grounds; observations for Lake Winnipegosis provided via personal communication by W. H. Koonz.

14. Estimates for population sizes are based on data in Wires et al. 2001 and Adkins and Roby 2010, a recent effort to compile the most current published and unpublished estimates from across the North American range for a fifth edition of the Wetlands International Waterbird Population Estimates, available at http://wpe.wetlands.org/, and unpublished data for Manitoba obtained mostly in 2012 from Scott Wilson (Canadian Wildlife Service).

15. Estimates available at http://wpe.wetlands.org/.

16. See Hatch and Weseloh 1999 for subspecies; Mercer 2008.

3. European Colonization and the Making of a Pariah

1. All historical references to the abundance of cormorants are included in Wires et al. 2001 and Wires and Cuthbert 2006 unless otherwise noted.

2. Cited in Mowat 1984.

3. Recorded in Lewis 1929.

4. Reported in Doughty 1975.

5. See Audubon 1835, "The Eggers of Labrador," and Mowat 1984 for particularly vivid accounts of egging.

6. All cited in Mowat 1984.

7. See Mowat 1984 and Montevecchi and Tuck 1987 for the demise of the great auk, and Montevecchi 2002 for the quoted comment.

8. Steller 1925; Hume and Walters 2012.

9. Jehl's revised estimate is noted in Hatch 1995.

10. Cited in Hatch 1995.

11. Records for New England are cited in Mendall 1936.

12. See Siegel-Causey et al. 1991 and Bailey 1993 for fur farming in Alaska and impacts on seabirds. See Ainley and Lewis 1974 and Mathewson 1986 for pressures that seabirds faced further south, and Grinnell 1928 and Carter et al. 1995 for Mexico.

13. Quoted in Mendall 1936.

14. Klingender 1953.

15. Sweetser 1867.

16. Noted in Wires et al. 2003.

17. Engels 1999; see Kors and Peters 2001 for other examples of black animals and evil associations.

18. See Howell 1928 and Knudsen 1951 for use of term "Nigger goose," and Finley 1915 for comments about Klamath Lake.

4. From Audubon to Conservation

1. Audubon 1843.

2. Ibid.

3. See Matthiessen 1987 for the development of wildlife laws.

4. Information about the development and history of the Audubon Society comes mostly from Graham 1990.

5. Noted in Merchant 2010.

6. Matthiessen 1987.

7. Recovery for New England after 1930 is summarized from Gross 1944 and Drury 1973–1974.

8. See Milton et al. 1995 for Nova Scotia, and Drury 1973–1974 for immigration as a factor in regional recovery.

9. Early records are cited in Baillie 1947 and Wires and Cuthbert 2006. See Postupalsky 1978 and Ludwig and Summer 1997 for population estimates.

10. Wires and Cuthbert 2006.

11. Wilson 1941 and Baillie 1947 report observations of "new" birds; Postupalsky 1978 and Weseloh and Collier 1995 describe "invasion."

12. See Wires et al. 2001 and Wires and Cuthbert 2006 for numerous examples.

13. Some additional records not discussed elsewhere include the 1821 and 1834 journals of the geographer and explorer Henry Schoolcraft, and the *History of the Ojibway People* (1855) by the historian William Warren. These records document cormorants at lakes in central Minnesota, about 100 miles west of Lake Superior, between 1745 and the early 1800s. Additionally, the use of cormorant place names by the Ojibwe can be traced at some of these locations through Warren's work. The unpublished field ledgers of L. Kumlien provide an additional, interesting record of a cormorant egg collected in 1870 at Lake Koshkonong, Wisconsin, about 55 miles west of Lake Michigan (Sumner Matteson, personal communication with the author). Nesting in the state was not formerly documented until about 1921, by which time cormorants had already declined from Lake Koshkonong and elsewhere.

14. Lewis 1929.

15. See McLeod and Bondar 1953 for Lake Winnipegosis, and Baillie 1947 for Lake of the Woods.

16. Bailey 1993.

17. Mathewson 1986.

18. Ainley and Lewis 1974.

5. Reversal of Fortunes

1. Events are summarized from McLeod and Bondar 1953.

2. McLeod and Bondar 1953.

3. See Gross 1951, Dow 1953, and Drury 1973–1974 for the history of Maine's control program, and Dow 1953 for an assessment of its effectiveness.

4. Baillie 1947; Postupalsky 1978.

5. DesGranges and Reed 1981.

6. Gress et al. 1973.

7. Drury 1973–1974.

8. Vermeer 1969; Vermeer and Peakall 1977; Hobson et al. 1989.

9. Anderson and Hickey 1975; Vermeer and Peakall 1977; Ludwig et al. 1995; Weseloh and Collier 1995; Weseloh et al. 1995.

10. Guillaumet et al. 2011.

11. Numbers on the Mississippi are reported in Audubon's journal of his trip to New Orleans (1820–1821), Grassett 1926, and Kirsch 1997; declines are reported in Anderson and Hammerstrom 1967.

12. See King et al. 1977 and Schreiber 1980 for information on brown pelican declines and endrin; details on endrin and its probable effect on cormorants provided by a personal communication from Ian Nisbet.

13. Anderson and Keith 1980; Velarde and Anderson 1990; Carter et al. 1995.

14. Anderson and Hamerstrom 1967.

15. Matteson et al. 1999.

16. Carson 1966.

17. Vermeer and Rankin 1984; Chapdelaine and Brousseau 1991.

18. Ludwig 1984; Weseloh et al. 1995. Information on commercial licenses was obtained by personal communications from J. P. Ludwig and D. V. Weseloh. Ludwig also provided information on reduced interactions between cormorants and fishermen.

19. Chapdelaine and Bédard 1995; Milton et al. 1995; Krohn et al. 1995; Wires et al. 2001.

20. Koonz and Rakowski 1985; Hobson et al. 1989; Wires et al. 2001; Somers et al. 2010. Alberta numbers come from a 2006 Innovation Alberta interview with the biologist A. McGregor: http://www.innovationalberta.com/article.php?articleid=684.

21. For PCBs and deformities, see Weseloh and Collier 1995, Ludwig et al. 1995, and Dunn 1997. For vitamin D deficiency, see Kuiken et al. 1999.

22. The story of Cosmos was provided by J. P. Ludwig via personal communication.

23. Weseloh et al. 1995; Wires et al. 2001; Weseloh et al. 2002.

24. Wires et al. 2001; Heinrich 2008.

25. Jackson and Jackson 1995; Wires et al. 2001.

26. Carter et al. 1995; Wires et al. 2001; Adkins and Roby 2010.

27. Ludwig 1984; Ludwig et al. 1989.

28. Hobson et al. 1989; Chapdelaine and Bédard 1995; Rail and Chapdelaine 1998; McGregor 2013.

29. Glahn et al. 1995; Jackson and Jackson 1995; Thompson et al. 1995.

30. Craven and Lev 1987; Ludwig et al. 1989; Ludwig and Summer 1997.

31. See Aderman and Hill 1995, Hatch 1995, and Nisbet 1995 for the context of winter numbers, and Jackson and Jackson 1995 for CBC data.

6. Fish Ponds and Reservoirs

1. Glahn and Stickley 1995; Aderman and Hill 1995; Snipes et al. 2005; http://www .belzonims.com/oldindex.html.

2. Price and Nickum 1995; Tucker 2012; Hanson and Sites 2012.

3. Tucker 2012; Hanson and Sites 2012.

4. Glahn and Stickley 1995; Jackson and Jackson 1995.

5. Hodges 1989; Price and Nickum 1995; USDA 2011b.

6. See Tucker 2012 for review of catfish-production practices and issues.

7. See Glahn and Stickley 1995 and Jackson and Jackson 1995 for changes in winter range use; see Glahn et al. 1995 and Glahn et al. 1999 for information on diet and body mass.

8. Price and Nickum 1995.

9. Ibid.

10. Martin and Hanson 1966.

11. Information from Texas State Historical Association, *Texas Almanac*, "Lakes and Reservoirs," http://www.texasalmanac.com/topics/environment/lakes-and -reservoirs; Oklahoma Water Resources Board, "Lakes of Oklahoma," http://www

.owrb.ok.gov/news/publications/lok/lok.php; J. Wertz and L. Layden, "The Factors That Fueled Oklahoma's Golden Age of Reservoir Building," October 1, 2012, http://stateimpact.npr.org/oklahoma/2012/10/01/the-factors-that-fueled-oklahomas-golden-age-of-reservoir-building/.

12. Official "nuisance" status noted in *News OK*, "Cormorant Declared Nuisance by Officials," March 14, 1991, http://newsok.com/cormorant-declared-nuisance-by-officials/article/2350563. For Texas, see T. Maxwell, "Cormorant Control Is an Issue," *San Angelo Standard-Times*, April 8, 2007, http://www.gosanangelo.com/news/2007/apr/08/cormorant-control-issue/?print=1.

7. Animal Damage Control and the First Standing Depredation Order for Cormorants

1. The history of animal damage control summarized from Acord 1992, GAO 2001, and Hawthorne 2004 unless otherwise noted.

2. Doughty 1975.

3. See Lopez 1978 for impacts on wolves; Schmidt 1986 provides the earlier figure for coyotes; USDA/APHIS provides more recent data on number of coyotes killed: http://www.aphis.usda.gov/wildlife_damage/prog_data/prog_data_report.shtml.

4. The American Society of Mammalogists, cited in Edge c. 1934, 5 (emphasis in the original); the Animal Damage Control Act of 1931, cited in USFWS 1979, 2.

5. Edge 1934, 6.

6. Leopold et al. 1964, 26–27.

7. Cain et al. 1972.

8. The Animal Damage Control Policy Study Committee report of 1978 was not obtained, but its recommendations are discussed in Acord 1992, GAO 2001, and Hawthorne 2004.

9. In addition to the sources noted above, see Wade 1986 and the Thoreau Institute's audit of the ADC program in 1994: http://www.ti.org/adcreport.html.

10. Acord 1989.

11. Acord 1989; Trapp et al. 1995.

12. Noted in USFWS 1999b.

13. Stickley and Andrews 1989.

14. Trapp et al. 1995; USFWS 1997.

15. Mott and Boyd 1995; Glahn, Reinhold, et al. 2000.

16. See the studies in Nettleship and Duffy 1995, and especially those in Glahn and Brugger 1995.

17. Mott and Boyd 1995; Trapp et al. 1995, 226–27, 229–30.

18. USFWS 1997.

19. Comments on the proposed order are summarized in the final rule (USFWS 1998).

20. USFWS 1998.

8. Conflicts on the Breeding Grounds

1. Muter et al. 2009.

2. Wires and Cuthbert 2003.

3. Details for Maine are from Krohn et al. 1995.

4. The note on the program's end is from USFWS 1999a.

5. USFWS 1999a.

6. DesGranges and Reed 1981, 18.

7. Bédard et al. 1995, 1999, 153.

8. Bédard et al. 1999.

9. Summarized from Milton et al. 1995.

10. Milton et al. 1995, 94.

11. Documented in Cairns et al. 1998.

12. Summarized from Koonz and Rakowsi 1985, Koonz 1989, and Hobson et al. 1989.

13. Hobson et al. 1989.

14. Sheppard 1994–1995.

15. Information obtained through personal communications with W. H. Koonz.

16. Craven and Lev 1987; Matteson et al. 1999.

17. Weseloh and Struger 1985.

18. Associated Press 1987; Ludwig et al. 1989.

19. Summarized from Ewins and Weseloh 1994, USFWS 1999b, and Johnson et al. 2002.

20. USFWS 1999a, 1999b; Shieldcastle and Martin 1999; Jarvie et al. 1999.

21. Observations for West and Herbert Islands were obtained through personal communications with Jim Ludwig; Pigeon Island observations are from Ewins and Wesloh 1994.

22. USFWS 1999a, 1999b.

23. Lucchesi 1988; Diana et al. 1997.

24. Matteson et al. 1999.

25. See Tobin 1999 for specific studies.

26. Trapp et al. 1999.

27. Revkin 1998a.

28. Revkin 1998b.

29. NYSDEC 1999.

9. The Second Standing Depredation Order for Cormorants

Note: Details on House and Senate bills, resolutions, and court cases are not provided but can be accessed through searches of state legislatures and other sources.

1. See Bennett 2002b and Scrivener 2009 for apocalypse imagery; for the comparison to a nuclear blast, see http://www.wcax.com/story/18850666/the-mission-to-control-cormorants; see Muter et al. 2009 for the media summary.

2. Muter and Triezenberg 2011.

3. See http://www.legis.state.tx.us/BillLookup/History.aspx?LegSess=76R& Bill=HCR23 for the history of HCR 23.

4. USFWS 1999a.

5. USFWS 1999b; see McCullough and Mazzocchi 2012 for the Little Galloo population objective rationale.

6. Summarized from Revkin 1999.

7. Revkin 1999.

8. Lydecker 2000; Mcgrath 2003.

9. USFWS 1999a, 1999b.

10. Duerr 2007; McCullough and Mazzocchi 2012.

11. GLSFC 1999.

12. See Heinrich 2008 for Minnesota fisheries data; the Peterson quotation is in Associated Press 2000.

13. *Catfish Journal* 1999; Hutchinson 1999.

14. Reinhold et al. 1998.

15. Wires et al. 2001.

16. USFWS 2003a.

17. Bennett 2002a.

18. Glahn, Reinhold, et al. 2000 gives revised replacement costs; Glahn et al. 2002 is the published version of the 2000 presentation.

19. Glahn, Tobin, et al. 2000, 5.

20. Great Lakes Fishery Commission, 2001 Committee Resolutions, Resolution #3: Double-crested cormorants: http://www.glfc.org/staff/resol2001.htm.

21. Position statement summarized from Godwin et al. 2003; Booth's comments are in Bennett 2002a.

22. Position statement posted at http://joomla.wildlife.org/WildlifeDamage/index.php?option=com_content&task=view&id=184.

23. Bennett 2002a.

24. Moore 2010.

25. Bennett 2002a.

26. USFWS 2003a.

27. USFWS 2003b.

10. A Half Million and Counting

1. Terry Doyle, USFWS, provided information on survey versus producer data; see Bennett 2002a for Booth's comments.

2. USFWS 2009 reports on numbers and locations between 2004 and 2007. Minnesota data are from USFWS 2003a and unpublished data from the Minnesota Department of Natural Resources provided by Katie Haws. (The number of cormorants taken in Minnesota in 2002 was not available, so an estimate was based on the average from other years during this time period.)

3. USFWS 2003b.

4. Ibid.

5. USFWS 2009.

6. See Fund for Animals v. Norton, 365 F. Supp.2d 394 (S.D.N.Y. 2005).

7. Environmental assessments active for cormorants at the time of this writing can be obtained at http://www.aphis.usda.gov/regulations/ws/ws_nepa_environmental _documents.shtml#EAs. For the purposes of this book, they are identified as USDA 2011-MI, USDA 2010-VT, USDA 2009-WI, USDA 2006-OH, and USDA 2005-MN. Additionally, Michigan's 2004 EA and 2006 amendment are identified as USDA 2004-MI and USDA 2006-MI, and can be obtained from Michigan Wildlife Services.

8. USDA 2004-MI; USFWS 2009.

9. The press release, nuisance permit instructions, and application packet are available at http://www.tpwd.state.tx.us.

10. Part of the public record authorizing cormorant control under the nuisance permit, this letter was written by H. Dale Hall, regional USFWS director, New Mexico office, and provided upon request by T. Doyle, USFWS.

11. See USDA 2005-MN for Minnesota; see 2005 Wisconsin Act 287 for Wisconsin's mandate for cormorant control.

12. USDA 2006-OH; USDA 2006-MI.

13. Described in reports prepared for the USFWS by the Chippewa Ottawa Resource Authority, the Little Traverse Bay Bands of Odawa Indians, the Grand Traverse Band of Ottawa and Chippewa Indians, and the Bay Mills Indian Community.

14. USDA 2009-WI; USDA 2010-VT.

15. See USDA 2011-MI for Michigan. Information for Wisconsin comes from a 2011 report on annual control activities prepared by Wisconsin Wildlife Services.

16. Data on nests destroyed is from USFWS 2009 and annual reports submitted by agencies summarizing control activities.

17. Shwiff et al. 2009.

18. USFWS 2010 documents declining trends in fishing; see Mcgrath 2003 for the quotation.

11. Looking North to Canada

1. See Scrivener 2009, DeGeorgio 2008, and CBC News 2009 for media references.

2. Management summaries for Presqu'ile Provincial Park are in Thorndyke 2008 and annual reports available at http://www.ontarioparks.com/english/pres _planning.html. Management summaries for Lac La Biche are in Thorndyke 2008, McGregor 2012, and personal communications from the biologist A. McGregor at Alberta Sustainable Resource Development. Management summaries for Middle Island are in Thorndyke 2008 and annual reports prepared by staff at Point Pelee National Parks of Canada. Management summaries for Lac St. Pierre are in MRNF 2006 and personal communications from the wildlife biologists Pascale Dombrowski and Charles Maisonneuve at the Ministère des Ressources Naturelles et de la Faune, Quebec.

3. Ridgway et al. 2012.

4. Thorndyke 2008; Taylor et al. 2011. Planning documents are available at http://trca.on.ca/.

5. Population data are from Wires et al. 2001. Information on Manitoba comes from personal communication with W. H. Koonz.

6. Summarized from Milton et al. 1995; the policy still guides management today.

7. OMNR 2006.

8. Based on MRNF 2006 and personal communication with Charles Maisonneuve, MNRF.

9. Fisheries (Alberta) Amendment Act, 2002, Statutes of Alberta, Chapter 14; McGregor 2012.

10. Press releases posted by the Peaceful Parks Coalition (http://www.peaceful parks.org) track many of these activities; see also Presqu'ile Provincial Park planning documents.

11. Some of these activities are documented at the Peaceful Parks Coalition website. Additionally, information was obtained through personal communications with the activists Anna Maria Valastro, Peaceful Parks Coalition; Julie Woodyer, Zoocheck International; Liz White, Animal Alliance of Canada; and Barry Kent McKay.

12. Peaceful Parks Coalition press releases from May 29 and June 6 record these events. See Presqu'ile Provincial Park planning documents for the impact of protesters' activities.

13. A Peaceful Parks Coalition press release from September 29, 2006, details the illegal exclusion zone and arrests.

14. Information from reports prepared by Parks Canada/Point Pelee National Park.

15. See Ontario Parks planning documents for Presqu'ile Provincial Park and East Sister Island.

16. From Taylor et al. 2011 and Toronto and Region Conservation Authority documents at http://trca.on.ca/.

17. In 2002 and 2003, the Alberta Wilderness Association published articles in the *Wild Lands Advocate* opposing cormorant management at Lac La Biche; see http://albertawilderness.ca.

18. Based on information provided by A. McGregor through personal communications.

19. See documents on the websites of the Peaceful Parks Coalition and Cormorant Defenders International, especially for East Sister Island.

20. History and designations for Tommy Thompson Park from Taylor et al. 2011.

12. Untangling the Mysteries between Predator and Prey

1. See Rudstam et al. 2004 for discussion.

2. See Wires et al. 2003 for the definition and scale of impacts.

3. See Wires et al. 2003 for guidelines from workshop; see VanDeValk et al. 2002 and Rudstam et al. 2004 for Oneida Lake; Yodzis 2001.

4. See studies by Diana et al. 2006 and Rudstam et al. 2004.

5. See Diana et al. 2006, Eisenhower and Parish 2009, Johnson et al. 2010, De-Bruyne et al. 2012, and DeBruyne et al. 2013 for examples.

6. Summarized from reviews of diet-assessment methods by Carss et al. 1997, Wires et al. 2001, and studies cited therein.

7. Colin Grubel conducted captive studies as part of doctoral research at CUNY Graduate Center and Queens College and provided data through a personal communication. See Russell et al. 1995 for information on European shag.

8. See studies cited in Wires et al. 2001 for ratios of nonbreeding to breeding birds.

9. See studies cited in Wires et al. 2001. For more recent studies documenting variation or highlighting the need for continual site-specific information, see Fenech et al. 2004, Johnson et al. 2010, DeBruyne et al. 2012, and DeBruyne et al. 2013.

10. Madenjian and Gabrey 1995.

11. Burnett et al. 2002; Willberg et al. 2004. See information from the Minnesota Department of Natural Resources at http://www.michigan.gov/documents/dnr/LBDN_walleye_management_strategy-_Jul2012_FINAL_391363_7.pdf and http://www.dnr.state.mn.us/enforcement/op_squarehook_faq.html.

12. See Yodzis 2001 for modeling approaches; Hilborn and Walters 1992 for an overview of fish population dynamics, compensation, and additive effects on secondary prey; and Lantry et al. 2002, Rudstam et al. 2004, and Fielder et al. 2008 for additive effects through consumption of juvenile fish.

13. The lack of studies meeting key data requirements is noted in VanDeValk et al. 2002 and Rudstam et al. 2004.

14. Trapp et al. 1999.

15. Coleman 2009; DeBruyne 2013.

16. NYSDEC 1999; Burnett et al. 2002; Johnson et al. 2002; Lantry et al. 2002.

17. See Burnett et al. 2002 for illegal harvesting.

18. Burnett et al. 2002, 208.

19. Johnson et al. 2010 documents the cormorant diet shift to round goby; Lantry 2012 details changes in smallmouth bass. Additional information on limits to bass recovery and the ability to monitor perch obtained through a personal communication with J. Lantry, NYSDEC.

20. Diana at al. 1997; Belyea et al. 1999; Diana ct al. 2006.

21. See Dorr et al. 2010 for management summary in this region.

22. Fielder 2010a.

23. Fielder 2010b.

24. Observations on expectations for fishery recovery are summarized from USDA 2006-MI and information provided by Jim Diana through personal communication. See Fielder et al. 2007 and the Michigan Department of Natural Resources website for improvements in yellow perch reproductive success elsewhere.

13. Adaptive Management

1. Hillborn and Waters 1992; Williams et al. 2009; Williams 2011.

2. See also Williams et al. 2009 and Williams 2011.

3. See Ridgway et al. 2006 for density dependence on Lake Huron.

4. See Taylor and Dorr 2003, 43, for the quotation; "wildlife acceptance capacity" was introduced by Decker and Purdy 1988.

5. Each of the programs implemented was described as an adaptive management approach. See Coleman 2009 for eastern Lake Ontario and Oneida Lake, New York, and Fielder 2008 for Les Cheneaux Islands, Michigan.

6. Management is summarized in Coleman 2009 and DeBruyne et al. 2013.

7. See Jackson et al. 2012 for fishery and lake monitoring information.

8. Recommendations of Cornell Biological Field Station fisheries scientists in Jackson et al. 2012; see the *Oneida Lake Bulletin*, Spring 2012 (http://www.oneidalake association.org/OLA%20Bulletin.pdf) for the Oneida Lake Association's influence and efforts to restore federal control efforts.

9. Data from McCullough and Mazzocchi 2012.

10. See the data in ibid.

11. Management summarized from Dorr et al. 2010 and the Michigan Wildlife Services report summarizing control activities of 2012; population objective noted in Polk 2008; see Michigan's amended environmental assessment 2006, 18, for minimum colony sizes.

12. Polk 2008.

13. Raccoon introduction is documented in Dorr et al. 2010.

14. See Lounsbury 2008 for media coverage.

15. Seider 2003; Seefelt 2005; Seefelt and Gillingham 2006.

16. See VanGuilder and Seefelt 2013 for round goby in the cormorant diet.

17. Information obtained via a personal communication from Sumner Matteson, Wisconsin Department of Natural Resources.

18. Meadows 2007.

19. Information obtained via a personal communication from Jeffrey Pritzl, Wisconsin Department of Natural Resources.

20. Summarized from the 2005 environmental assessment unless otherwise noted.

21. Information on modeling, initial increases in walleye, and signs of recovery in the presence of larger cormorant numbers than targeted was obtained from earlier reports prepared by the Band and via a personal communication from Steve Mortensen, Leech Lake Band of Ojibwe.

22. See Mosedale 2008 for media coverage.

23. See USDA 2011-MI for monitoring limitations.

24. Data on Michigan's coastal area and inland waters were summarized from a presentation prepared by Michigan Sea Grant in 2012.

25. Ridgway et al. 2012.

26. USDA 2009-WI; USDA 2011-MI.

27. USDA 2009-WI; USDA 2011-MI.

14. Back to the Wintering Grounds

1. The minute order, a legal document recorded in the minutes of a court session, is noted in Atlantic and Mississippi Flyway Council 2010; additional information provided via personal communication from Karen Rowe, Arkansas Game and Fish Commission.

2. USDA 2011b.

3. Tucker 2012; USDA 2011b; see also http://msucares.com/aquaculture/catfish/disease.html.

4. Wywialowski 1999; Glahn, Reinhold, et al. 2000.

5. Glahn and Dorr 2002.

6. Glahn et al. 2002.

7. Glahn, Tobin, et al. 2000.

8. References include USFWS 2003a, 2003b, and USDA 2004; for the Bird Depredation Committee comments, see http://naturalresources.house.gov/uploadedfiles/baxter_6.24.04.pdf; see also Taylor and Dorr 2003; Werner and Dorr 2006; Taylor and Strickland 2008; King et al. 2010; Dorr et al. 2012a, 2012b.

9. Sullivan et al. 2006; National Sea Grant Law Center, http://nsglc.olemiss.edu/SandBar/SandBar4/4.2comorant.htm.

10. USFWS 2009.

11. Atlantic and Mississippi Flyway Council 2010.

12. USDA 2011a.

13. Dorr et al. 2012a.

14. Ibid.

15. Keller et al. 1998.

16. Conover et al. 2007.

17. Observations and data presented in this review are from a Texas Parks and Wildlife Department report summarizing its control activities under the Public Resource Depredation Order, September 2010 through August 2011. Additionally, numbers of cormorants killed between 2008 and 2011 in Texas counties under the order (via the state's nuisance permit) were data the agency provided to the USFWS as required under the order.

18. See Simmonds et al. 2000 for the Oklahoma study, and Campo et al. 1993 for the Texas study.

19. Montgomery 2010.

15. Engineer or Destroyer

1. CDI 2008.

2. For rates of deforestation, see the Food and Agriculture Association of the United Nations, *State of the World's Forests 2012*, http://www.fao.org/docrep/016/i3010e/i3010e00.htm; for Great Lakes island use by cormorants, see Wires and Cuthbert 2010.

3. For the Middle Island Conservation Plan, see http://www.pc.gc.ca/pn-np/on/pelee/plan/plan1.aspx.

4. See Presqu'ile Provincial Park Planning Documents, especially the Recommendations Report from November 2004, http://www.ontarioparks.com/english/pres_planning.html, and USDA 2006-OH.

5. See Vigmostad 1999 for detail on the dynamic nature of islands; see Ludwig 1974 for waterbird use cycles.

6. Night-heron information from Quilliam 1965; gull information obtained from Great Lakes colonial waterbird survey monitoring efforts.

7. Wires and Cuthbert 2010.

8. See Ricciardi 2001 for numbers of exotic species in the Great Lakes.

9. Duncan et al. 2011.

10. Cuthbert et al. 2002.

11. See Wires et al. 2010 for species priorities and conservation recommendations; quotations are from p. 82.

12. Summarized from Weseloh et al. 1988, Dawson 1903, and Peterjohn 1989.

13. USFWS 2000.

14. See Baillie 1947 for the first Lake Erie breeding records for cormorants; see Peterjohn 1989 for occurrence of the great egret.

15. See Weseloh et al. 1988, 1995, for the cormorant's colonization history and persistence in western Lake Erie.

16. History of Middle Island summarized from Weseloh et al. 1988 and CDI 2008.

16. Opening Pandora's Box

1. See the Wilderness Act of 1964, Public Law 88-577 (16 U.S.C. 1131–1136).

2. See USDA 2006-OH for USFWS's "non-degradation principle" and justification for controlling cormorants at National Wilderness Areas.

3. For the Michigan Nature Association's position on cormorants, see http://michigannature.wordpress.com/2012/04/06/protecting-the-cormorants; for the reported 500-yard zone and data on birds shot via "off colony shooting," see USDA 2011-MI; some information was compiled from Wildlife Services' report to the USFWS for take in 2011 under the Public Resource Depredation Order.

4. For conservation goals, see http://www.partnersinflight.org, http://www.waterbirdconservation.org, and http://www.birdlife.org; see also Wires and Cuthbert 2006.

5. Wires and Cuthbert 2001.

6. Mosedale 2008.

7. USFWS 2003b; Atlantic and Mississippi Flyway Council 2010.

8. Information from Steve Mortenson obtained through personal communication; debate over waterfowl crippling rates communicated via personal communication from the waterfowl biologist Todd Arnold, University of Minnesota.

9. Observations for Gull and Whiskey Islands provided via personal communication from the wildlife scientist Nancy Seefelt, Central Michigan University.

10. See environmental assessments for Ohio, Michigan, and Wisconsin (USDA 2006-OH; USDA 2009-WI; USDA 2011-MI) for identical text on humane issues.

11. See http://www.aphis.usda.gov/wildlife_damage/annual%20tables/00table 10t.pdf for 2000; see http://www.aphis.usda.gov/wildlife_damage/prog_data/ prog_data_report_FY2008.shtml for information on blackbirds.

12. The American Society of Mammalogists' letter is posted on the society's website at http://www.mammalsociety.org/uploads/committee_files/ASM-Federal%20 wildlife%20control%20letter_0.pdf; information on the lawsuit brought by WildEarth Guardians is at http://www.wildearthguardians.org.

13. Reed et al. 2003.

References

In writing this book, I consulted numerous published and unpublished sources. Many of the published references are available in four comprehensive sources: publications of symposia by Nettleship and Duffy 1995 and Tobin 1999; the North American Status Assessment for double-crested cormorants by Wires et al. 2001; and a review of current and historical populations of cormorants by Wires and Cuthbert 2006. Most of the numerous historical records consulted are not included here but are available in the collections by Wires et al. 2001 and Wires and Cuthbert 2006, and information is provided in the chapters to identify the records used. Additionally, the Notes section identifies works that contain multiple sources relevant to particular chapters. For other sources cited but not listed here, there is generally enough information provided to identify the work or the source can be easily found (for example, Shakespeare's plays). Most of the government documents referenced are available online, and most unpublished reports can be obtained by contacting the appropriate agencies. The complete bibliography is available at the Wetlands International-Cormorant Research Group website, http://cormorants.freehostia.com/index.htm.

Acord, B. R. 1989. The USDA-APHIS-ADC program in the United States. *Fourth Eastern Wildlife Damage Control Conference*. Ed. Scott R. Craven. Madison, WI.

———. 1992. Responses of the ADC program to a changing American society. *Proceedings of the Fifteenth Vertebrate Pest Conference*. Ed. J. E. Borrecco and R. E. Marsh. Davis: University of California, 1992.

Aderman, A. R., and E. P. Hill. 1995. Locations and numbers of double-crested cormorants using winter roosts in the Delta region of Mississippi. In Nettleship and Duffy 1995, 143–151.

Adkins, J. Y., and D. D. Roby. 2010. *A Status Assessment of the Double-Crested Cormorant (Phalacrocorax auritus) in Western North America: 1998–2009*. Final Report to the US Army Corps of Engineers, Portland, OR.

Ainley, D. G., and T. J. Lewis. 1974. The history of Farallon Island marine bird populations, 1854–1972. *Condor* 76: 432–446.

Anderson, D. W., and F. Hamerstrom. 1967. The recent status of Wisconsin cormorants. *Passenger Pigeon* 29 (1): 3–15.

Anderson, D. W., and J. J. Hickey. 1972. Eggshell changes in certain North American birds. *Proceedings of the International Ornithological Congress* 15: 514–540.

Anderson, D. W., and J. O. Keith. 1980. The human influence on seabird nesting success: Conservation implications. *Biological Conservation* 18: 65–80.

Arnott, W. G. 2007. *Birds in the Ancient World from A to Z.* New York: Routledge.

Ashmole, N. P. 1963. The regulation of numbers of tropical oceanic birds. *Ibis* 103b: 458–473.

Associated Press. 1987. Cormorant eggs found destroyed. May 30.

———. 2000. Representative seeks season to hunt for cormorant birds. June 8.

Atlantic and Mississippi Flyway Council. 2010. *Atlantic and Mississippi Flyways Double-Crested Cormorant Management Plan.* Prepared by the Cormorant Ad hoc Committees, Atlantic and Mississippi Flyway Council, Nongame Migratory Bird Technical Section. March.

Audubon, J. J. 1835. *Ornithological Biography.* Edinburgh. 3: 420–425.

———. 1843. *Birds of America,* vol. 6. Reprint, New York: Dover, 1967.

Bailey, E. P. 1993. Introduction of foxes to Alaskan Islands: History, effects on avifauna, and eradication. US Department of the Interior, Fish and Wildlife Service, Resources, Publication 193. Washington, DC: Government Printing Office.

Baillie, J. L. 1947. The double-crested cormorant nesting in Ontario. *Canadian Field-Naturalist* 61: 119–126.

Bédard, J., A. Nadeau, and M. Lepage. 1995. Double-crested cormorant culling in the St. Lawrence River Estuary. In Nettleship and Duffy 1995, 78–85.

———. 1999. Double-crested cormorant culling in the St. Lawrence River Estuary: Results of a five-year program. In Tobin 1999, 147–154.

Belyea, G. Y., et al. 1999. Impact of double-crested cormorant predation on the yellow perch population in Les Cheneaux islands of Michigan. In Tobin 1999, 47–59.

Bennett, D. 2002a. Agencies differ on cormorant control. *Delta Farm Press* (Clarksdale, MS), Mar. 8.

———. 2002b. The cormorant apocalypse is upon us. *Delta Farm Press* (Clarksdale, MS), Nov. 8.

Bergstrom, B. J., et al. 2013. License to kill: Reforming federal wildlife control to restore biodiversity and ecosystem function. *Conservation Letters.* First published online, July 11, 2013.

Birt, V. L., et al. 1987. Ashmole's halo: Direct evidence for prey depletion by a seabird. *Marine Ecology—Progress Series* 40: 205–208.

Bovy, K. M. 2011. Archaeological evidence for a double-crested cormorant (*Phalacrocorax auritus*) colony in the Pacific Northwest, USA. *Waterbirds* 34 (1): 89–95.

Brachet, A. 1878. *An Etymological Dictionary of the French Language*. 3rd ed. Trans. G. W. Kitchin. Oxford: Clarendon Press.

Brodkorb, P. 1963. Catalogue of fossil birds, Part 1: Archaeopterygiformes through Ardeiformes. *Bulletin of the Florida State Museum: Biological Sciences* 7 (4): 179–293.

Bur, M. T., S. L. Tinnirello, C. D. Lovell, and J. T. Tyson. 1999. Diet of the double-crested cormorant in western Lake Erie. In Tobin 1999, 73–85.

Burnett, J. A. D., N. H. Ringler, B. F. Lantry, and J. H. Johnson. 2002. Double-crested cormorant predation on yellow perch in the eastern basin of Lake Ontario. *Journal of Great Lakes Research* 28: 202–211.

Cain, S. A., et al. 1972. *Predator Control, 1971: A Report to the President's Council on Environmental Quality and the U.S. Department of the Interior by the Advisory Committee on Predator Control*. Ann Arbor: University of Michigan Press.

Cairns, D. K., and J. J. Kerekes. 2000. Fish harvest by common loons and common mergansers in Kejimkujik National Park, Nova Scotia, Canada, as estimated by bioenergetic modeling. In *Limnology and Aquatic Birds: Monitoring, Modeling, and Management*, ed. F. A. Comin, J. A. Herrera, and J. Ramirez, 125–135. Mérida, Mexico: Universidad Autónoma de Yucatán.

Cairns, D. K., R. L. Dibblee, and P. Y. Daoust. 1998. Displacement of a large double-crested cormorant, *Phalacrocorax auritus*, colony following human disturbance. *Canadian Field-Naturalist* 112: 520–522.

Campo, J. J., et al. 1993. Diet of double-crested cormorants wintering in Texas. *Journal of Field Ornithology* 64: 135–144.

Carson, R. 1962. *Silent Spring*. Boston: Houghton Mifflin.

Carson, R. D. 1966. Destruction of colonial birds on an island on Suggi Lake. *Blue Jay* 24: 96–97.

Carss, D. N., et al. 1997. Techniques for assessing cormorant diet and food intake: Towards a consensus view. *Supplemento Ricerche di Biologia della Selvaggina* 26: 197–230.

Carter, H. R., et al. 1995. Population size, trends, and conservation problems of the double-crested cormorant on the Pacific Coast of North America. In Nettleship and Duffy 1995, 189–215.

Catfish Journal. 1999. Cormorant nesting area located in southwest Arkansas. Vol. 8 (11): 14–15.

CBC News. 2009. Smelly cormorants overtake old Hillsborough piers. CBC News, Prince Edward Island. Sept. 1.

CDI [Cormorant Defenders International]. 2008. *Observations of the 2008 cormorant cull on Middle Island in Point Pelee National Park*. September.

Chapdelaine, G., and J. Bédard. 1995. Recent changes in the abundance and distribution of double crested cormorants in the St. Lawrence River, Estuary, and Gulf, Québec. 1978–1990. In Nettleship and Duffy 1995, 70–77.

Chapdelaine, G., and P. Brousseau. 1991. Thirteenth census of seabird populations in the sanctuaries of the North Shore of the Gulf of St. Lawrence, 1982–1988. *Canadian Field-Naturalist* 105: 60–66.

Chesser, R. T., et al. 2010. Fifty-first supplement to the American Ornithologists' Union check-list of North American birds. *Auk* 127: 726–744.

Coleman, J. T. H. 2009. Diving behavior, predator-prey dynamics, and management efficacy of double-crested cormorants in New York State. PhD diss., Cornell University.

Conover, G., R. Simmonds, and M. Whalen, eds. 2007. Management and control plan for bighead, black, grass, and silver carps in the United States. Asian Carp Working Group. Washington, DC: Aquatic Nuisance Species Task Force.

CPR [Center for Progressive Reform]. 2011. Making good use of adaptive management. White Paper 1104. http://www.progressivereform.org/whitePapers .cfm. Accessed May 2013.

Craven, S. R., and E. Lev. 1987. Double-crested cormorants in the Apostle Islands, Wisconsin, USA: Population trends, food habits, and fishery depredations. *Colonial Waterbirds* 10: 64–71.

Cuthbert, F. J., L. R. Wires, and J. E. McKearnan. 2002. Potential impacts of nesting double-crested cormorants on great blue herons and black-crowned night-herons in the U.S. Great Lakes region. *Journal of Great Lakes Research* 28: 145–154.

Dalton, C. M., D. Ellis, and D. M. Post. 2009. The impact of double-crested cormorant (*Phalacrocorax auritus*) predation on anadromous alewife (*Alosa pseudoharengus*) in south-central Connecticut, USA. *Canadian Journal of Fisheries and Aquatic Sciences* 66: 177–186.

Dawson, W. L. 1903. *The Birds of Ohio*. Columbus, OH: Wheaton.

DeBruyne, R. L., et al. 2012. Spatial and temporal comparisons of double-crested cormorant diets following the establishment of alewife in Lake Champlain, USA. *Journal of Great Lakes Research* 38: 123–130.

———. 2013. Analysis of prey selection by double-crested cormorants: A 15-year diet study in Oneida Lake, New York. *Transactions of the American Fisheries Society* 142: 430–446.

Decker, D. J., and K. G. Purdy. 1988. Toward a concept of wildlife acceptance capacity in wildlife management. *Wildlife Society Bulletin* 16: 53–57.

DeGeorgio, L. 2008. Cormorants running amok. *My Town Crier* (Toronto). July 16.

DesGranges, J., and A. Reed. 1981. Disturbance and control of selected colonies of double-crested cormorants in Quebec. *Colonial Waterbirds* 4: 12–19.

Diana, J. S. 2010. Should cormorants be controlled to enhance yellow perch in Les Cheneaux Islands? A comment on Fielder (2008). *Journal of Great Lakes Research* 36: 190–194.

Diana, J. S., G. Y. Belyea, and R. D. Clark, Jr., eds. 1997. History, status, and trends in populations of yellow perch and double-crested cormorants in Les Cheneaux

Islands, Michigan. Michigan Department of Natural Resoures, Fisheries Division, Special Report 16. Ann Arbor.

Diana, J. S., S. Maruca, and B. Low. 2006. Do increasing cormorant populations threaten sportfishes in the Great Lakes? A case study in Lake Huron. *Journal of Great Lakes Research* 32: 306–320.

Dorr, B. S., T. Aderman, P. H. Butchko, and S. C. Barras. 2010. Management effects on breeding and foraging numbers and movements of double-crested cormorants in the Les Cheneaux Islands, Lake Huron, Michigan. *Journal of Great Lakes Research* 36: 224–231.

Dorr, B. S., L. W. Burger, S. C. Barras, and K. C. Godwin. 2012a. Economic impact of double-crested cormorant, *Phalacrocorax auritus*, depredation on channel catfish, *Ictalurus punctatus*, aquaculture in Mississippi, USA. *Journal of the World Aquaculture Society* 43: 502–513.

————. 2012b. Double-crested cormorant distribution on catfish aquaculture in the Yazoo River Basin of Mississippi. *Wildlife Society Bulletin* 36: 70–77.

Doucette, J., B. Wissel, and C. M. Sommers. 2011. Cormorant-fisheries conflicts: Stable isotopes reveal a consistent niche for avian piscivores in diverse food webs. *Ecological Applications* 21: 2987–3001.

Doughty, R. W. 1975. *Feather Fashions and Bird Preservation*. Berkeley: University of California Press.

Dow, R. L. 1953. *The Herring Gull–Cormorant Control Program: State of Maine*. State of Maine Department of Sea and Shore Fisheries, General Bulletin No. 1. Augusta.

Drury, W. H. 1973–1974. Population changes in New England seabirds. *Bird Banding* 44 (4): 267–313, 45 (1): 1–15.

Duerr, A. 2007. Population dynamics, foraging ecology, and optimal management of double-crested cormorants on Lake Champlain. PhD diss., University of Vermont.

Duncan, T., et al. 2011. Flora of the Erie islands: A review of the floristic, ecological, and historical research and conservation activities, 1976–2010. *Ohio Journal of Science* 110: 3–12.

Dunn, K. 1997. Drastic Deformities. Frontline: Fooling with Nature. www.pbs.org/wgbh/pages/frontline/shows/nature/gallery/cormorants.html.

Edge, R. c. 1934. *The United States Bureau of Destruction and Extermination: The Misnamed and Perverted "Biological Survey."* Rosalie Barrow Edge Manuscript Collection, Denver, CO.

Eisenhower, M. D., and D. L. Parrish. 2009. Double-crested cormorant and fish interactions in a shallow basin of Lake Champlain. *Waterbirds* 32: 388–399.

Engels, D. 1999. *Classical Cats: The Rise and Fall of the Sacred Cat*. London: Routledge.

Enstipp, M. R., R. D. Andrews, and D. R. Jones. 2001. The effects of depth on the cardiac and behavioural responses of double-crested cormorants (*Phalacrocorax auritus*) during voluntary diving. *Journal of Experimental Biology* 204: 4081–4092.

Ewins, P. J., and D. V. C. Weseloh. 1994. Effects on productivity of shooting double-crested cormorants (*Phalacrocorax auritus*) on Pigeon Island, Lake Ontario, in 1993. *Journal of Great Lakes Research* 20: 761–767.

Feduccia, A. 1980. *The Age of Birds*. Cambridge, MA: Harvard University Press.

———. 1999. *The Origin and Evolution of Birds*. 2nd ed. New Haven: Yale University Press.

Fenech, A . S., S. E. Lochman, and A. A. Radomski. 2004. Seasonal diets of male and female double-crested cormorants from an oxbow lake in Arkansas, USA. *Waterbirds* 27: 170–176.

Fielder, D. G., 2004. Collapse of the yellow perch fishery in Les Cheneaux Islands, Lake Huron, and possible causes. *Proceeding of Percis III: The Third International Percid Fish Symposium*, ed. T. P. Barry and J. A. Malison, 129–130. University of Wisconsin Sea Grant Institute, Madison.

Fielder, D. G., J. S. Schaeffer, and M. V. Thomas. 2007. Environmental and ecological conditions surrounding the production of large year classes of walleye (*Sander vitreus*) in Saginaw Bay, Lake Huron. *Journal of Great Lakes Research* 33 (SP1): 118–132.

———. 2008. Examination of factors contributing to the decline of the yellow perch population and fishery in Les Cheneaux Islands, Lake Huron, with emphasis on the role of double-crested cormorants. *Journal of Great Lakes Research* 34: 506–523.

———. 2010a. Response to Diana commentary. *Journal of Great Lakes Research* 36: 195–198.

———. 2010b. Response of yellow perch in Les Cheneaux Islands, Lake Huron to declining numbers of double-crested cormorants stemming from control activities. *Journal of Great Lakes Research* 36: 207–214.

Finley, W. L. 1915. Cruising the Klamath. *Bird-Lore* 17: 485–401.

Fitzpatrick, J. W. 2002. The AOU and bird conservation: Recommitment to the revolution. *Auk* 119: 907–913.

GAO [General Accounting Office]. 2001. *Wildlife Services Program: Information on Activities to Manage Wildlife Damage*. Report to Congressional Committees, GAO-02-138.

Glahn, J. F., and K. E. Brugger. 1995. The impact of double-crested cormorants on the Mississippi Delta catfish industry: A bioenergetics model. In Nettleship and Duffy 1995, 168–175.

Glahn, J. F., and B. S. Dorr. 2002. Captive double-crested cormorant *Phalacrocorax auritus* predation on channel catfish *Ictalurus punctutus* fingerlings and its influence on single-batch cropping production. *Journal of the World Aquaculture Society* 33: 85–93.

Glahn, J. F., and A. R. Stickley, Jr. 1995. Wintering double-crested cormorants in the Delta region of Mississippi: Population levels and their impact on the catfish industry. In Nettleship and Duffy 1995, 137–142.

Glahn, J. F., J. B. Harrel, and C. Vyles. 1998. The diet of wintering double-crested cormorants feeding at lakes in the southeastern United States. *Colonial Waterbirds* 21: 446–452.

Glahn, J. F., D. S. Reinhold, and C. A. Sloan. 2000. Recent population trends of double-crested cormorants wintering in the Delta region of Mississippi: Responses to roost dispersal and removal under a recent depredation order. *Waterbirds* 23: 34–44.

Glahn, J. F., M. E. Tobin, and B. F. Blackwell. 2000. *A Science-Based Initiative to Manage Double-Crested Cormorant Damage to Southern Aquaculture.* US Department of Agriculture/APHIS/WS 11–55–010.

Glahn, J. F., M. E. Tobin, and J. B. Harrel. 1999. Possible effects of catfish exploitation on overwinter body condition of double-crested cormorants. In Tobin 1999, 147–154.

Glahn, J. F., P. J. Dixon, G. A. Littauer, and R. B. McCoy. 1995. Food habits of double-crested cormorants wintering in the Delta region of Mississippi. In Nettleship and Duffy 1995, 158–167.

Glahn, J. F., S. J. Werner, T. Hanson, and C. R. Engle. 2002. Cormorant depredation losses and their prevention at catfish farms: Economic considerations. In *Proceedings of the Third NWRC Special Symposium, "Human Conflicts with Wildlife: Economic Considerations,"* ed. L. Clark.

GLSFC [Great Lakes Sport Fishing Council]. 1999. Bill introduced to control/hunt cormorants. *Weekly News, Fishery News of the Great Lakes Basin.* Dec. 6.

Godwin, K. C., T. King, P. Butchko, and R. Chapman. 2003. Challenges of implementing the double-crested cormorant environmental impact statement. USDA National Wildlife Research Center, Staff Publications, Paper 238. In *Proceedings of the 10th Wildlife Damage Management Conference,* ed. K. A. Fagerstone and G. W. Witmer.

Graham, F., Jr. 1990. *The Audubon Ark: A History of the National Audubon Society.* New York: Knopf.

Grassett, F. 1926. An unusual flight of cormorants. *Wilson Bulletin* 38 (4): 234–235.

Grémillet, D. 1995. "Wing-drying" in cormorants. *Journal of Avian Biology* 26 (2): 176.

Grémillet, D., R. P. Wilson, S. Wanless, and G. Peters. 1999. A tropical bird in the Arctic (the cormorant paradox). *Marine Ecology Progress Series* 188: 305–309.

Grémillet, D., et al. 2004. Linking the foraging performance of a marine predator to local prey abundance. *Functional Ecology* 18: 793–801.

———. 2005. Cormorants dive through the Polar night. *Biology Letters* 1: 469–471.

Gress, F., et al. 1973. Reproductive failures of double-crested cormorants in southern California and Baja California. *Wilson Bulletin* 85: 197–208.

Grinnell, J. 1928. A distributional summation of the ornithology of Lower California. *University of California Publications in Zoology* 32: 1–300.

Gross, A. O. 1944. The present status of the double-crested cormorant on the coast of Maine. *Auk* 61 (4): 513–537.

———. 1951. The herring gull–cormorant control project. *Proceedings of the 10th International Ornithological Congress*, ed. Sven Hörstadius, 532–536. Stockholm: Almqvist and Wiksell.

Guillaumet, A., et al. 2011. Determinants of local and migratory movements of Great Lakes double-crested cormorants. *Behavioral Ecology* 22: 1096–1103.

Hanson, T., and D. Sites. 2012. *2011 U.S. catfish database*. Mississippi State University, Department of Agricultural Economics, Information Report 2012-01.

Hatch, J. J. 1982. The cormorants of Boston Harbor and Massachusetts Bay. *Bird Observer of Eastern Massachusetts* 10 (2): 65–73.

———. 1995. Changing populations of double-crested cormorants. In Nettleship and Duffy 1995, 8–24.

Hatch, J. J., and D. V. Weseloh. 1999. Double-crested cormorant (*Phalacrocorax auritus*). In *The Birds of North America*, no. 441, ed. A. Poole and F. Gill. Philadelphia: Birds of North America.

Hawthorne, D. W. 2004. The history of federal and cooperative animal damage control. *Sheep and Goat Research Journal* 19: 13–15.

Hebert, C. E., D. V. Weseloh, E. M. Senese, and G. D. Haffner. 2005. Unique island habitats may be threatened by double-crested cormorants. *Journal of Wildlife Management* 69 (1): 68–76.

Heinrich, T. 2008. Population trends of walleye, sauger, yellow perch, and double-crested cormorants on Lake of the Woods, Minnesota. Report to Minnesota Department of Natural Resources.

Hillborn, R., and C. J. Walters. 1992. *Quantitative Fisheries Stock Assessment: Choice, Dynamics and Uncertainty*. New York: Chapman and Hall.

Hobson, K. A., and J. C. Driver. 1989. Archaeological evidence for the use of the Straight of Georgia by marine birds. In *The Ecology and Status of Marine and Shoreline Birds in the Straight of Georgia, British Columbia*, ed. K. Vermeer and R.W. Butler, 168–173. Canadian Wildlife Service Special Publication, Ottawa.

Hobson, K. A., R. W. Knapton, and W. Lysack. 1989. Population, diet, and reproductive success of double-crested cormorants breeding on Lake Winnipegosis, Manitoba, in 1987. *Colonial Waterbirds* 12: 191–197.

Hodges, M. F. 1989. Foraging by piscivorous birds on commercial fish farms in Mississippi. MSc thesis, Mississippi State University.

Howell, A. H. 1928. *The Birds of Alabama*. Department of Game and Fisheries of Alabama. Montgomery.

Hume, J. P., and M. Walters. 2012. *Extinct Birds*. London: Bloomsbury.

Hutchinson, J. 1999. Large cormorant nesting site found in Arkansas. *Aquaculture News* 7 (9): 1–2.

Jackson, J. A., and B. J. S. Jackson. 1995. The double-crested cormorant in the south-central United States: Habitat and population changes of a feathered pariah. In Nettleship and Duffy 1995, 118–130.

Jackson, J. R., et al. 2012. The fisheries and limnology of Oneida Lake, 2000–2011. New York Federal Aid in Sport Fish Restoration. Study 2, F-61-R.

Jarvie, S., H. Blokpoel, and T. Chipperfield. 1999. A geographic information system to monitor nest distributions of double-crested cormorants and black-crowned night-herons at shared colony sites near Toronto, Canada. In Tobin 1999, 121–129.

Johnsgard, P. A. 1993. *Cormorants, Darters, and Pelicans of the World.* Washington, DC: Smithsonian Institution Press.

Johnson, J. H., R. M. Ross, and R. D. McCullough. 2002. Little Galloo Island, Lake Ontario: A review of nine years of double-crested cormorant diet and fish consumption information. *Journal of Great Lakes Research* 28: 182–192.

Johnson, J. H., R. M. Ross, R. D. McCullough, and A. Mathers. 2010. Diet shift of double-crested cormorants in eastern Lake Ontario associated with the expansion of the invasive round goby. *Journal of Great Lakes Research* 36: 242–247.

Keller, T., A. von Lindeiner, and U. Lanz. 1998. Cormorant management in Bavaria, Southern Germany: Shooting as a proper management tool? *Cormorant Research Group Bulletin* 3: 11–15.

King, D. T., J. F. Glahn, and K. J. Andrews. 1995. Daily activity budgets and movements of winter roosting double-crested cormorants determined by biotelemetry in the Delta region of Mississippi. In Nettleship and Duffy 1995, 152–157.

King, D. T., B. F. Blackwell, B. S. Dorr, and J. L. Belant. 2010. Effects of aquaculture on migration and movement patterns of double-crested cormorants. *Human-Wildlife Conflicts* 4: 77–86.

King, K. A., E. L. Flickinger, and H. H. Hildebrand. 1977. The decline of brown pelicans on the Louisiana and Texas Gulf Coast. *Southwestern Naturalist* 21: 417–431.

King, R. J. 2013. *The Devil's Cormorant: A Natural History.* Durham: University of New Hampshire Press.

Kirsch, E. M. 1995. Double-crested cormorants along the Upper Mississippi River. In Nettleship and Duffy 1995, 131–136.

———. 1997. Numbers and distribution of double-crested cormorants on the Upper Mississippi River. *Colonial Waterbirds* 20: 177–184.

Klingender, F. D. 1953. St. Francis and the birds of the apocalypse. *Journal of the Warburg and Courtauld Institutes* 16 (1–2): 13–23.

Knudsen, G. J. 1951. An interesting Wisconsin rookery. *Passenger Pigeon* 8: 119–124.

Koonz, W. H. 1989. Double-crested cormorant populations and food habits on Lake Winnipegosis, Manitoba. Unpublished report. Wildlife Branch, Manitoba Department of Natural Resources, Winnipeg.

Koonz, W. H., and P. W. Rakowski. 1985. Status of colonial waterbirds nesting in southern Manitoba. *Canadian Field-Naturalist* 99 (1): 19–29.

Kors, A. C., and E. Peters. 2001. *Witchcraft in Europe, 400–1700: A Documentary History.* 2nd ed. Philadelphia: University of Pennsylvania Press.

Krohn, W. B., R. B. Allen, J. R. Moring, and A. E. Hutchinson. 1995. Double-crested cormorants in New England: Population and management histories. In Nettleship and Duffy 1995, 99–109.

Kuiken, T., F. A. Leighton, G. Wobeser, and B. Wagner. 1999. Causes of morbidity and mortality and their effect on reproductive success in double-crested cormorants from Saskatchewan. *Journal of Wildlife Diseases* 35: 331–346

Lantry, B. F., T. H. Eckert, C. P. Schneider, and J. R. Chrisman. 2002. The relationship between the abundance of smallmouth bass and double-crested cormorants in the Eastern Basin of Lake Ontario. *Journal of Great Lakes Research* 28: 193–201.

Lantry, J. R. 2012. Eastern Basin of Lake Ontario warmwater fisheries assessment, 1976–2011. In *NYSDEC Lake Ontario Annual Report 2011*, sec. 4, 1–24. Albany: New York State Department of Environmental Conservation, Bureau of Fisheries.

Leopold, A. S., et al. 1964. *Predator and Rodent Control in the United States: Report Submitted to Department of Interior* ["Leopold Report"].

Lewis, H. F. 1929. *The Natural History of the Double-Crested Cormorant (*Phalacrocorax auritus auritus*) [Lesson])*. Ottawa: Ru-Mi-Lou Books.

Lewis, S., T. N. Sherratt, K. C. Hamer, and S. Wanless. 2001. Evidence of intra-specific competition for food in a pelagic seabird. *Nature* (London) 412: 816–819.

Lopez, B. H. 1978. *Of Wolves and Men*. New York: Scribner's Sons.

Lounsbury, H. 2008. Wildlife managers thinning cormorant flocks near Oscoda, U.P. *Bay City (MI) Times*, May 7.

Lucchesi, D. O. 1988. A biological analysis of the yellow perch population in the Les Cheneaux Islands, Lake Huron. MS thesis, University of Michigan.

Ludwig, J. P. 1974. Recent changes in the ring-billed gull population and biology in the Laurentian Great Lakes. *Auk* 91: 575–594.

———. 1984. Decline, resurgence, and population dynamics of Michigan and Great Lakes double-crested cormorants. *Jack Pine Warbler* 62 (4): 91–102.

Ludwig, J. P., and C. L. Summer. 1997. Population status and diet of cormorants in Les Cheneaux Islands area. In *History, Status, and Trends in Populations of Yellow Perch and Double-Crested Cormorants in Les Cheneaux Islands, Michigan*, ed. J. S. Diana, G. Y. Belyea, and R. D. Clark, Jr., 5–25. Special Report 16. Ann Arbor: Fisheries Division, Michigan Department of Natural Resources.

Ludwig, J. P., C. N. Hull, M. E. Ludwig, and H. J. Auman. 1989. Food habits and feeding ecology of nesting double-crested cormorants in the upper Great Lakes, 1986–1989. *Jack Pine Warbler* 67 (4): 115–126.

Ludwig, J. P., et al. 1995. Evaluation of the effects of toxic chemicals in Great Lakes cormorants: Has causality been established? In Nettleship and Duffy 1995, 60–69.

Lydecker, R. 2000. Game fish controversy takes wing; cormorants killed in New York to save bass fish. *Boat/US*, Mar.

Lysack, W. 2006. The Lake Winnipegosis commercial fishery monitoring program, 1990–2005. MS-Report no. 2006-01. Manitoba Water Stewardship Fisheries.

Mackay, G. H. 1894. Habits of the double-crested cormorant (*Phalacrocorax dilophus*) in Rhode Island. *Auk* 11: 18–25.

Madenjian, C. P., and S. W. Gabrey. 1995. Waterbird predation on fish in western Lake Erie: A bioenergetics model application. *Condor* 97: 141–153.

Martin, R. O. R., and R. L. Hanson. 1966. *Reservoirs in the United States.* US Geological Survey Water-Supply Paper 1838. Washington, DC: Government Printing Office.

Marzluff, J. M., and T. Angell. 2005. *In the Company of Crows and Ravens.* New Haven: Yale University Press.

Mathewson, W. 1986. *William L. Finley: Pioneer Wildlife Photographer.* Corvallis: Oregon State University Press.

Matteson, S. W., et al. 1999. Changes in the status, distribution, and management of double-crested cormorants in Wisconsin. In Tobin 1999, 27–45.

Matthiessen, P. 1987. *Wildlife in America.* New York: Viking.

McCullough, R., and I. Mazzocchi. 2012. Cormorant management activities in Lake Ontario's Eastern Basin. In *NYSDEC Lake Ontario Annual Report 2011,* sec. 13, 1–8. Albany: New York State Department of Environmental Conservation, Bureau of Fisheries.

Mcgrath, S. 2003. Shoot-out at Little Galloo. *Smithsonian,* Feb., 72–78.

McGregor, A. M. 2012. *Double-Crested Cormorant (Phalacrocorax auritus) Monitoring and Management in Lac La Biche, Alberta: 2011 Annual Summary.* Edmonton: Alberta Sustainable Resource Development, Fish and Wildlife Division.

———. 2013. Fish harvest and replacement of top piscivorous predators in aquatic food webs: Implications for restoration and fisheries management. PhD diss., University of Alberta, Edmonton.

McLeod, J. A., and G. F. Bondar. 1953. A brief study of the double-crested cormorant on Lake Winnipegosis. *Canadian Field-Naturalist* 67 (1): 1–11.

Meadows, S. A. 2007. Food habits of double-crested cormorants in southern Green Bay with emphasis on impacts on the yellow perch fishery: Final summary report prepared for the Wisconsin Department of Natural Resources.

Mendall, H. L. 1936. The home-life and economic status of the double-crested cormorant, *Phalacrocorax auritus auritus* (Lesson). *Maine Bulletin* 39 (3): 1–159.

Mercer, D. M. 2008. Phylogeography and population genetic structure of double-crested cormorants (*Phalacrocorax auritus*). MS thesis, Oregon State University, Corvallis.

Merchant, C. 2010. George Bird Grinnell's Audubon Society: Bridging the gender divide in conservation. *Environmental History* 15: 3–30.

Milton, G. R., P. J. Austin-Smith, and G. J. Farmer. 1995. Shouting at shags: A case study of cormorant management in Nova Scotia. In Nettleship and Duffy 1995, 91–98.

Minteer, B. A., and J. P. Collins. 2005. Ecological ethics: Building a new tool kit for ecologists and biodiversity managers. *Conservation Biology* 19: 1803–1812.

Montevecchi, W. A. 2002. Review of "Who Killed the Great Auk?" *Auk* 119 (4): 1211–1213.

Montevecchi, W. A., and L. M. Tuck. 1987. *Newfoundland Birds: Exploitation, Study, Conservation.* Cambridge, MA: Nuttall Ornithological Club.

Montgomery, R. 2010. No magic bullet for cormorant control. *Bassmaster,* Oct 21.

Moore, K. 2010. Agency profile. *Arkansas Agriculture* 7 (4): 24.

Mosedale, M. 2008. Most hated bird in the world: Sanctioned killing of cormorants continues unabated in Minnesota. *MinnPost,* July 16.

Mott, D. F., and F. L. Boyd. 1995. A review of techniques for preventing double-crested cormorant depredations at aquaculture facilities in the Southeastern U.S. In Nettleship and Duffy 1995, 176–180.

Mowat, F. 1984. *Sea of Slaughter.* Boston: Atlantic Monthly Press.

MRNF [Ministère des Ressources Naturelles et de la Faune]. 2006. *Increase in Québec's Double-Crested Cormorant Population: Should We Be Concerned?* English translation, Aug.

Muter, B. A., and H. A. Triezenberg. 2011. More than social media: Social networks in wildlife management. *Wildlife Professional* 2011 (Winter): 70–72.

Muter, B. A., M. L. Gore, and S. J. Riley. 2009. From victim to perpetrator: Evolution of risk frames related to human-cormorant conflict in the Great Lakes. *Human Dimensions of Wildlife* 14: 366–379.

Nettleship, D. N., and D. C. Duffy. 1995. The double-crested cormorant: Biology, conservation and management. Special publication 1, *Colonial Waterbirds* 18.

Nisbet, I. C. T. 1995. Biology, conservation, and management of the double-crested cormorant: Symposium summary and overview. In Nettleship and Duffy 1995, 247–252.

Norton, A. H., and R. P. Allen. 1931. Breeding of the great black-backed gull and double-crested cormorant in Maine. *Auk* 48: 589–592.

NYSDEC [New York State Department of Environmental Conservation]. 1999. *Final Report: To Assess the Impact of Double-Crested Cormorant Predation on the Smallmouth Bass and Other Fishes of the Eastern Basin of Lake Ontario.* Albany: New York State Department of Environmental Conservation, Bureau of Fisheries; Washington, DC: US Geological Survey, Biological Resources Division.

OMNR [Ontario Ministry of Natural Resources]. 2006. *Review of the Status and Management of Double-Crested Cormorants in Ontario, 2006.* Peterborough: Ontario Ministry of Natural Resources, Fish and Wildlife Branch, Wildlife Section.

Peterjohn, B. 1989. *The Birds of Ohio.* Bloomington: Indiana University Press.

Polk, A. 2008. Biologists, sportsmen outline cormorant control plans. *St. Ignace (MI) News,* Mar. 27.

Pollard, J. 1977. *Birds in Greek Life and Myth.* London: Thames and Hudson.

Postupalsky, S. 1978. *Toxic Chemicals and Cormorant Populations in the Great Lakes.* Manuscript Report no. 40. Ottawa: Canadian Wildlife Service, Wildlife Toxicology Division.

Price, I. M., and J. G. Nickum. 1995. Aquaculture and birds: The context for contro-versy. In Nettleship and Duffy 1995, 33–45.

Quilliam, H. R. 1965. *The History of the Birds of Kingston, Ontario.* Kingston: Privately printed.

Rail, J., and G. Chapdelaine. 1998. Food of double-crested cormorants, *Phalacrocorax auritus*, in the gulf and estuary of the St. Lawrence River, Quebec, Canada. *Canadian Journal of Zoology* 76: 635–643.

Reed, J. M., et al. 2003. Review of the double-crested cormorant management plan, 2003: Final report of the AOU conservation committee's panel.

Reinhold, D. S., A. J. Mueller, and G. Ellis. 1998. Observations of nesting double-crested cormorants in the Delta region of Mississippi. *Colonial Waterbirds* 21 (3): 466–467.

Revkin, A. C. 1998a. A slaughter of cormorants in angler country. *New York Times*, Aug. 1.

———. 1998b. In fishing hamlet, no grief cormorants; near the scene of a slaughter, the protected birds are blamed for a decrease in jobs and revenue. *New York Times*, Aug. 9.

———. 1999. Nine men plead guilty to slaughtering cormorants to protect sport fish-ing. *New York Times*, Apr. 9.

Ricciardi, A. 2001. Facilitative interactions among aquatic invaders: Is an "invasional meltdown" occurring in the Great Lakes? *Canadian Journal of Fisheries and Aquatic Sciences* 58: 2513–2525.

Ridgway, M. S. 2010. Seasonal and annual patterns in density of double-crested cor-morants in two coastal regions of Lake Huron. *Journal of Great Lakes Re-search* 36: 411–418.

Ridgway, M. S., T. A. Middel, and J. B. Pollard. 2012. Response of double-crested cormorants to a large-scale egg oiling experiment on Lake Huron. *Journal of Wildlife Management* 76: 740–749.

Ridgway, M. S., J. B. Pollard, and D. V. Weseloh. 2006. Density-dependent growth of double-crested cormorant colonies on Lake Huron. *Canadian Journal of Zoology* 84: 1409–1420.

Rippey, E., J. J. Rippey, and J. N. Dunlop. 2002. Increasing numbers of pied cormo-rants breeding on the islands off Perth, Western Australia and the conse-quences for the vegetation. *Corella* 26: 61–64.

Rudstam, L. G., et al. 2004. Cormorant predation and the population dynamics of wall-eye and yellow perch in Oneida Lake. *Ecological Applications* 14: 149–163.

Russell, A. F., S. Wanless, and M. P. Harris. 1995. Factors affecting the production of pellets by shags, *Phalacrocorax aristotelis*. *Seabird* 17: 44–49.

Schmidt, R. H. 1986. Community-level effects of coyote population reduction. In *Community Toxicity Testing*, ed. J. Cairns, Jr., 49–65. Philadelphia: Amer-ican Society for Testing and Materials.

Schreiber, R. W. 1980. The brown pelican: An endangered species? *BioScience* 30: 743–747.

Scrivener, L. 2009. 30,000 cormorants destroying lakeside park. *Toronto Star*, May 20.

Seefelt, N. E. 2005. Foraging ecology, bioenergetics, and predatory impact of breeding double-crested cormorants (*Phalacrocorax auritus*) in the Beaver Archipelago, northern Lake Michigan. PhD diss., Michigan State University.

Seefelt, N. E., and J. C. Gillingham. 2006. Foraging locations of double-crested cormorants in the Beaver Archipelago of northern Lake Michigan: Implications for smallmouth bass declines. *Waterbirds* 29: 473–480.

———. 2008. Bioenergetics and prey consumption of breeding double-crested cormorants in the Beaver Archipelago, northern Lake Michigan. *Journal of Great Lakes Research* 34: 122–133.

Seider, M. J. 2003. Population dynamics of smallmouth bass in the Beaver Archipelago, northern Lake Michigan, 1999–2002. MS thesis, University of Georgia.

Sellers, R. M. 1995. Wing-spreading behavior of the cormorant, *Phalacrocorax carbo*. *Ardea* 83: 27–36.

Sheppard, Y. 1994–1995. Cormorants and pelicans: Scapegoats for a fishing industry gone bad. *Birds of the Wild* 3 (4): 30–35.

Shieldcastle, M. C., and L. Martin. 1999. Colonial waterbird nesting on West Sister Island National Wildlife Refuge and the arrival of double-crested cormorants. In Tobin 1999, 115–119.

Shwiff, S., K. Kirkpatrick, and T. DeVault. 2009. *The Economic Impact of Double-Crested Cormorants to Central New York*. Fort Collins, CO: National Wildlife Research Center.

Siegel-Causey, D. 1988. Phylogeny of the Phalacrocoracidae. *Condor* 90: 885–905.

———. 1999. The problems of being successful: Managing interactions between humans and double-crested cormorants. In Tobin 1999, 5–16.

Siegel-Causey, D., D. Lefevre and A. B. Savinetskii. 1991. Historical diversity of cormorants and shags from Amchitka Island, Alaska. *Condor* 93: 840–852.

Simmonds, R. L., Jr., A. V. Zale, and D. M. Leslie, Jr. 2000. Modeled effects of double-crested cormorant predation on simulated reservoir sport and forage fish populations in Oklahoma. *North American Journal of Fisheries Management* 20: 180–191.

Snipes, C. E., et al. 2005. *Current Agricultural Practices of the Mississippi Delta*. Bulletin 1143. Mississippi State University, Mississippi Agricultural and Forestry Experiment Station.

Somers, C. M., V. A. Kjoss, F. A. Leighton, and D. Fransden. 2010. American white pelicans and double-crested cormorants in Saskatchewan: Population trends over five decades. *Blue Jay* 68: 75–86.

Steller, G. W. 1925. *Steller's Journal of the Sea Voyage from Kamchatka to America and Return on the Second Expedition, 1741–1742*. Trans L. Stejneger. In *Bering's Voyages: An Account of the Efforts of the Russians to Determine the Relation of Asia and America*, ed. F. A. Golder, vol. 2. New York: American Geographical Society.

Stickley, A. R., Jr., and K. J. Andrews. 1989. Survey of Mississippi catfish farmers on means, effort, and costs to repel fish-eating birds from ponds. In *Proceedings: Fourth Eastern Wildlife Damage Control Conference*, ed. S. R. Craven, 105–108. Madison: University of Wisconsin Cooperative Extension Service.

Stickley, A. R., Jr., G. L. Warrick, and J. F. Glahn. 1992. Impact of double-crested cormorant depredations on channel catfish farms. *Journal of the World Aquaculture Society* 23 (3): 192–198.

Strod, T., Z. Arad, I. Izhaki, and G. Katzir. 2004. Cormorants keep their power: Visual resolution in a pursuit diving bird under amphibious and turbid conditions. *Current Biology* 14: R376–R377.

Sullivan, K. L., P. D. Curtis, R. B. Chipman, and R. D. McCullough. 2006. *The Double-Crested Cormorant: Issues and Management*. Ithaca, NY: Cornell University Cooperative Extension Service.

Sweetser, P. H. 1867. The Bible birds (X). *Ladies Repository* 37: 372–378.

Taylor, B., D. Andrews, and G. S. Fraser. 2011. Double-crested cormorants and urban wilderness: Conflicts and management. *Urban Ecosystems* 14: 377–394.

Taylor, J., and B. Strickland. 2008. Effects of roost shooting on double-crested cormorant use of catfish ponds: Preliminary results. In *Proceedings of the Twenty-Third Vertebrate Pest Conference*, ed. R. M. Timm and M. B. Madon, 98–102. Davis: University of California.

Taylor, J. D., and B. S. Dorr. 2003. Double-crested cormorant impacts to commercial and natural resources. In *Proceedings of the Tenth Wildlife Damage Management Conference*, ed. K. A. Fagerstone and G. W. Witmer, 43–51. Fort Collins, CO: National Wildlife Research Center.

Thompson, B. C., J. J. Campo, and R. C. Telfair. 1995. Origin, population attributes, and management conflict resolution for double-crested cormorants wintering in Texas. In Nettleship and Duffy 1995, 181–188.

Thorndyke, R. 2008. *Overview of the Management of Double-Crested Cormorants (Phalacrocorax auritius): Canada and the United States*. Ottawa: Parks Canada.

Tobin, M. E. 1999. *Symposium on Double-Crested Cormorants: Population Status and Management Issues in the Midwest*. Technical Bulletin no. 1879. Washington, DC: US Department of Agriculture, Animal and Plant Health Inspection Service.

Trapp, J. L., S. J. Lewis, D. M. Pence. 1999. Double-crested cormorant impacts on sport fish: Literature review, agency survey, and strategies. In Tobin 1999, 87–96.

Trapp, J. L., T. J. Dwyer, J. J. Doggett, and J. G. Nickum. 1995. Management responsibilities and policies for cormorants: United States Fish and Wildlife Service. In Nettleship and Duffy 1995, 226–230.

Tucker, C. 2012. Channel catfish. In *Aquaculture Farming Aquatic Animals and Plants*, ed. J. S. Lucas and P. C. Southgate, 365–385. Oxford: Wiley-Blackwell.

Tudge, C. 2008. *The Bird*. New York: Three Rivers.

USDA [US Department of Agriculture]. 2004. *Defining Economic Impacts and Developing Strategies for Reducing Avian Predation in Aquaculture Systems.* Fort Collins, CO: US Department of Agriculture, Animal and Plant Health Inspection Service.

———. 2011a. *Protecting Agriculture.* Fort Collins, CO: US Department of Agriculture, Animal and Plant Health Inspection Service, Wildlife Services, National Wildlife Research Center.

———. 2011b. *Catfish 2010, Part 3: Changes in Catfish Health and Production Practices in the United States, 2002–09.* Fort Collins, CO: US Department of Agriculture, Animal and Plant Health Inspection Service, Veterinary Services, National Animal Health Monitoring System.

USFWS [US Fish and Wildlife Service]. 1979. *Mammalian Predator Damage Management for Livestock Protection in the Western United States: Final Environmental Impact Statement.* Washington, DC: US Department of the Interior.

———. 1995. *Cormorants and Their Impacts on Fish.* Washington, DC: US Department of the Interior, Fish and Wildlife Service, Division of Migratory Bird Management.

———. 1997. *Migratory Bird Permits: Proposed Depredation Order for the Double-Crested Cormorant; Proposed Rule.* Federal Register 62 (120): 33,960–33,965 (June 23).

———. 1998. *Migratory Bird Permits: Establishment of a Depredation Order for the Double-Crested Cormorant; Final Rule.* Federal Register 63 (42): 10,550–10,561.

———. 1999a. *Final Environmental Assessment of a USFWS Action to Issue a Migratory Bird Depredation Permit for the Take of Cormorants and Gulls on Lake Champlain Islands, Vermont.* May 3.

———. 1999b. *Final Environmental Assessment of a USFWS Action to Issue a Migratory Bird Depredation Permit for the Take of Cormorants on Lake Ontario Islands, New York.* May 3.

———. 2000. *Ottawa National Wildlife Refuge Complex Comprehensive Conservation Plan.*

———. 2003a. *Final Environmental Impact Statement: Double-Crested Cormorant Management in the United States.* Washington, DC: US Department of the Interior, Fish and Wildlife Service. http://www.fws.gov/migratorybirds/CurrentBirdIssues/Management/cormorant/CormorantFEIS.pdf.

———. 2003b. *Migratory Bird Permits: Regulations for Double-Crested Cormorant Management; Final Rule.* Federal Register 68 (195): 58,022–58,037 (Oct. 8).

———. 2009. *Final Environmental Assessment: Extended Management of Double-Crested Cormorants under 50 CFR 21.47 and 21.48.* Arlington, VA: US Department of the Interior, Fish and Wildlife Service, Division of Migratory Bird Management.

———. 2010. *Trends in Fishing and Hunting, 1991–2006: A Focus on Fishing and Hunting by Species; Addendum to the 2006 National Survey of Fishing, Hunting, and Wildlife-Associated Recreation.* Report 2006–8. Washington, DC: US Department of the Interior, Fish and Wildlife Service.

VanDeValk, A., et al. 2002. Comparison of angler and cormorant harvest of walleye and yellow perch in Oneida Lake, New York. *Transactions of the American Fisheries Society* 131: 27–39.

Van Guilder, M. A., and N. E. Seefelt. 2013. Double-crested cormorant (*Phalacrocorax auritus*) chick bioenergetics following round goby (*Neogobius melanostomus*) invasion and implementation of cormorant population control. *Journal of Great Lakes Research* 39: 153–161.

Van Tets, G. F. 1976. Australasia and the origin of shags and cormorants, Phalacrocoracidae. *Proceedings of the Sixteenth International Ornithological Congress*, 121–124. Canberra: Australian Academy of Science.

Velarde, E., and D. W. Anderson. 1990. Conservation and management of seabird islands in the Gulf of California: Setbacks and successes. In *Seabirds on Islands: Threats, Case Studies, and Action Plans*, ed. D. N. Nettleship, J. Burger, and M. Gochfeld, 229–243. Cambridge: BirdLife International.

Vermeer, K. 1969. The present status of double-crested cormorant colonies in Manitoba. *Blue Jay* 27: 217–220.

Vermeer, K., and D. B. Peakall. 1977. Toxic chemicals in Canadian fish-eating birds. *Marine Pollution Bulletin* 8: 205–210.

Vermeer, K., and L. Rankin. 1984. Population trends in nesting double-crested and pelagic cormorants in Canada. *Murrelet* 65: 1–9.

Vigmostad, K. E., ed. 1999. *State of the Great Lakes Islands: Proceedings from the 1996 U.S.-Canada Great Lakes Islands Workshop*. East Lansing: Michigan State University, Department of Resource Development.

Von Brandt, A. 2005. *Von Brandt's Fish Catching Methods of the World*. 4th ed. Ed. O. Gabriel. Oxford: Blackwell.

Wade, D. A. 1986. Predator damage control, 1980 to 1986. In *Proceedings of the Twelfth Vertebrate Pest Conference*, ed. T. P. Salmon, 369–386. Davis: University of California.

Warburton, B., and B. G. Norton. 2009. Towards a knowledge-based ethic for lethal control of nuisance wildlife. *Journal of Wildlife Management* 73: 158–164.

Werner, S. J., and B. S. Dorr. 2006. Influence of fish stocking density on the foraging behavior of double-crested cormorants *Phalacrocorax auritus*. *Journal of the World Aquaculture Society* 37: 121–125.

Weseloh, D. V., and B. Collier. 1995. *The Rise of the Double-Crested Cormorant on the Great Lakes: Winning the War against Contaminants*. Great Lakes Fact Sheet. Burlington, ON: Environment Canada.

Weseloh, D. V., and J. Struger. 1985. Massive mortality of juvenile double-crested cormorants on Little Galloo Island, July 1984. *Kingbird* 35: 98–104.

Weseloh, D. V., S. M. Teeple, and H. B. Blokpoel. 1988. The distribution and status of colonial waterbirds nesting in western Lake Erie. In *The Biogeography of the*

Island Region of Western Lake Erie, ed. J. F. Downhower, 134–144. Columbus: Ohio State University Press.

Weseloh, D. V., et al. 1995. Double-crested cormorants of the Great Lakes: Changes in population size, breeding distribution and reproductive output between 1913 and 1991. In Nettleship and Duffy 1995, 48–59.

———. 2002. Population trends and colony locations of double-crested cormorants in the Canadian Great Lakes and immediately adjacent areas, 1990–2000: A manager's guide. *Journal of Great Lakes Research* 28: 125–144.

White, C. R., P. J. Butler, D. Grémillet, and G. R. Martin. 2008. Behavioural strategies of cormorants (Phalacrocoracidae) foraging under challenging light conditions. *Ibis* 150 (S1): 231–239.

White, C. R., N. Day, P. J. Butler, and G. R. Martin. 2007. Vision and foraging in cormorants: More like herons than hawks? *PLoS ONE* 2 (7): e639. doi:10.1371/journal.pone.0000639.

Wild, D. 2012. *The Double-Crested Cormorant: Symbol of Ecological Conflict*. Ann Arbor: University of Michigan Press.

Willberg, M., J. Bence, and D. F. Clapp. 2004. *Development of an Age-Structured Yellow Perch Population Model for Lake Michigan: Final Report for Great Lakes Fish and Wildlife Restoration Act Project*. Ann Arbor, MI.

Williams, B. K. 2011. Adaptive management of natural resources: Framework and issues. *Journal of Environmental Management* 92: 1346–1353.

Williams, B. K., R. C. Szaro, and C. D. Shapiro. 2009. *Adaptive Management: The U.S. Department of the Interior Technical Guide*. Washington, DC: US Department of the Interior, Adaptive Management Working Group.

Wilson, H. 1941. By the wayside. *Passenger Pigeon* 3: 11.

Wires, L. R., and F. J. Cuthbert. 2001. *Prioritization of Waterbird Colony Sites for Conservation in the U.S. Great Lakes: Final Report to U.S. Fish and Wildlife Service*. Fort Snelling, MN: US Fish and Wildlife Service.

———. 2003. *Fish-Eating Bird Predation at Aquaculture Facilities in Minnesota: A First Step towards Bridging the Information Gap*. Final report to Minnesota Sea Grant. Duluth: Minnesota Sea Grant.

———. 2006. Historic populations of the double-crested cormorant (*Phalacrocorax auritus*): Implications for conservation and management in the 21st century. *Waterbirds* 29: 9–37.

———. 2010. Characteristics of double-crested cormorant colonies in the U.S. Great Lakes island landscape. *Journal of Great Lakes Research* 36: 232–241.

Wires, L. R., D. N. Carss, F. J. Cuthbert, and J. J. Hatch. 2003. Transcontinental connections in relation to cormorant-fisheries conflicts: Perceptions and realities of a "bête noire" (black beast) on both sides of the Atlantic. *Vogelwelt* 124 (S): 389–400.

Wires, L. R., F. J. Cuthbert, D. R. Trexel, and A. R. Joshi. 2001. *Status of the Double-Crested Cormorant (Phalacrocorax auritus) in North America: Final Report to U.S. Fish and Wildlife Service*. St. Paul: University of Minnesota, Department of Fisheries and Wildlife. http://digitalmedia.fws.gov/cdm/ref/collection/document/id/1430.

Wires, L. R., et al. 2010. *Upper Mississippi Valley / Great Lakes Waterbird Conservation Plan: Final Report*. Fort Snelling, MN: US Fish and Wildlife Service.

Wywialowski, A. P. 1999. Wildlife-caused losses for producers of channel catfish *Ictalurus punctatus* in 1996. *Journal of the World Aquaculture Society* 30: 461–472.

Yodzis, P. 2001. Must top predators be culled for the sake of fisheries? *Trends in Ecology and Evolution* 16: 78–84.

Index

Page numbers in *italics* refer to illustrations.